BEFORE THE WALL FELL:

THE SCIENCE OF MAN IN SOCIALIST EUROPE

EDITED BY
BECKY A. SIGMON
UNIVERSITY OF TORONTO

CANADIAN SCHOLARS' PRESS INC.   TORONTO   1993

Before the Wall Fell: The Science of Man in Socialist Europe

First published in 1993 by
Canadian Scholars' Press Inc.
402-180 Bloor St. W.
Toronto, ON M5S 2V6

Q
127
.E852
B43
1993

Canadian Cataloguing in Publication Data

Main entry under title:

Before the Wall Fell: The Science of Man in Socialist Europe

Includes bibliographical references.
ISBN 1-55130-010-9

1. Science and state - Europe, Eastern. I. Sigmon, Becky A.
(Becky Ann), 1941 - .

Q127.E82B45   1993   509.43   C92-095540-I

Canadian Scholars' Press Inc. wishes to acknowledge that the
idea for the cover came from a postcard the editor found in East
Berlin in 1987. The postcard was titled "Haus am Checkpoint
Charlie" by the artist, Siegfried Rischar.

27067280

Acknowledgements

This book is the creation of many people, and my role has been primarily to act as the catalyst to bring it all together. First and foremost, there are the European participants to thank for travelling to Toronto in order to share their research and their thoughts with us. Each one is representative of many more like them in their countries. In Toronto there are many of my friends and colleagues who contributed their time, homes, cars and a variety of other energies to making it possible to host our European participants; these are David Begun, Carole Fisher, Am Grewal, Ruby and Glenn Mack, Isabel Pegg, and Loren Vanderlinden. I am especially grateful for the support given me by Jack McCaffrey, my daughter Lisel Sigmon Storck and my son Jeffrey Ludlow Storck. My colleagues, Peter Storck and Lisa Golombek, at the Royal Ontario Museum, made it possible for the participants to tour the research facilities there. The Chairman of the Anthropology Department at the University of Toronto, Gary Crawford, arranged a special luncheon for departmental members and participants which resulted in a valuable interchange for all concerned.

For assistance in preparation of the manuscripts for publication, I am particularly grateful to my colleagues, Hermann Helmuth, Nancy Ossenberg and Emöke Szathmary, who have provided invaluable comments on the papers. Helmuth and Szathmary spent part of their childhood under socialist dominated countries, respectively of East Germany and Hungary, and both have a special kind of understanding for the circumstances under which their European colleagues have practised their profession.

The Ministry of External Affairs of Canada, especially the Toronto branch under Mary Harker, was extremely helpful both in giving advice about participant travel, and in getting air tickets to certain of our colleagues, for which I would like to express my appreciation.

The Symposium was funded by grants from the Natural Sciences and Engineering Research Council of Canada, and the Wenner-Gren Foundation for Anthropological Research in New York. Further grant funds were made available by the Natural Sciences and Engineering Research Council for publication of the symposium proceedings.

Finally, I would like to thank Valentyna Leonovičová, my co-organizer of the Symposium, for all the time, energy and creativity that she has given to the making of this book. It has been a privilege knowing her both as a colleague and as a friend.

THIS BOOK WAS PUBLISHED WITH THE
AID OF A GRANT FROM THE
NATURAL SCIENCES AND ENGINEERING
RESEARCH COUNCIL OF CANADA

TABLE OF CONTENTS

FOREWORD

Science, to me, must be and is, an international, all-human endeavor. If and when it becomes isolated and restricted by cultural, political and religious barriers, it suffers, becoming incomplete, distorted and fragmented. No longer able to answer the questions which it asks, it is incapable of pursuing the truth. Many of us have seen this happen in this century, but also, we have been fortunate to see truth prevail. The '90s are, I hope, an example of the latter. When the Berlin Wall was built in 1961, and after the invasion of Czechoslovakia in 1967, cultural-scientific relations between countries of the Warsaw Pact and NATO deteriorated to unprecedentedly low levels. Exchanges in science and of literature and scientists were nearly prohibited by an official or unofficial, but in any case, very efficient political embargo. The free flow of scientific information by means of periodicals, books and conference participation became very limited; long-term cooperation between scientists on projects of mutual interest and benefit in Physical Anthropology, were halted. The effects were an isolation of both sides which is still with us today. Though the West claims that it is advanced due to modern competition, it is wrong to assume that the East has stood still in its research. True, in some areas of Physical Anthropology such as Genetics, the problem of Race and the study of Human Evolution, scientists in Eastern Bloc countries were under very strong political pressure to conform to the current ideology. But other areas such as Growth, Anthropometry and Evolution flourished. Whereas Physical Anthropology in the West could afford to be more esoteric and pursue "pure knowledge", our colleagues in the East had to work hard to be useful and to apply themselves and their science for the benefit of the people and the working class. Adapting to the political demands, they worked in the fields of Growth and Development, Sports Anthropometry, the Secular Trend, Clinical Anthropometry, or even in Human Evolution, and made advances in areas where we neither felt interest nor pressure to work.

We in the West need to be informed of their current advances and vice versa. I may actually claim that the loss of scientific information was greater on our side than on theirs. Many of us send reprints, journals or books to our colleagues in the East. All too often, their financial constraints and our linguistic isolation prevented us from reading their information, but they were almost always able to read other languages such as English, French or German. In many areas, this complete ignorance or disregard of the achievements of our colleagues in the former East Germany, CSSR, Poland, Hungary, the former USSR and particularly Albania, is serious. I must admit that even I, being originally German, did not know all of the important sites of Early Hominid culture in the former GDR, let alone Albania. But our scientific world view must be universal and as thorough as possible. We need information on fossil sites in Albania as much as information on the famous site of Olduvai in order to build or refute theories. Otherwise we are not true to our declared goal: the pursuit of knowledge and truth. Will there be a dawning of a new age of Enlightenment? I sincerely hope so: I believe we are already on our way considering the important steps which have been taken over the last few years in speaking to each other.

This book marks an important progress in this direction. We need to find out what our colleagues have done in the last 30 years or more. We also have to catch up and read what our colleagues have written. It allows us to gain a historical perspective on the pursuit of Biological Anthropology which will, hopefully, be of benefit to our science and promote cooperation today and in the future. Maybe we can prevent past errors from happening again?

I am very glad that my colleague, Professor Becky Sigmon undertook the hard work to invite us to participate in her Symposium on Physical Anthropology in the Eastern Bloc countries and that she is publishing the papers for the common good of everybody. I am thankful to my colleagues in the former East Bloc countries for making their science and their advances known to us, and I hope we can work together towards our common goal to answer the eternal questions: What are we? Where do we come from? Where do we go from here?

Hermann Helmuth
Department of Anthropology, Trent University
Peterborough, Ontario
November 11, 1992

CONTRIBUTORS

MARINA L. BUTOVSKAYA is a Primatologist at the Institute of Ethnography, Adademy of Sciences, Moscow, Russia. Her primary interest is in studying primate behaviour as a means of understanding human behavioural evolution. Her research has been carried out primarily at the Sukumi Primate Research Center in Georgia. She has published articles focussing on primate group behaviour for interpreting human social behaviour, the role of aggression in primates, infanticide, the influence of kinship and personality in social and sexual behaviour, and the role of social factors in human evolution, to list a few areas of her interests. Whenever possible, she travels outside of Russia to participate in conferences and to present reports on her own research. Her work in primate behaviour is insightful and contributes interesting ideas to the literature in this field.

MICHAEL H. CRAWFORD is Director of the Laboratory of Biological Anthropology, University of Kansas, USA. He has carried out field work in anthropological genetics on a number of peoples in the world including, for example, the Tlaxcaltecans of Mexico, Black Caribs, Irish, Mennonites, Newfoundlanders, and the Tiszahats of Hungary. His research in Siberian anthropological genetics is related to the more general question of Arctic population biology and peopling of the New World. Crawford has both edited and co-edited several books on anthropological genetics including 3 volumes on *Current Developments in Anthropological Genetics*. He has published extensively on his research, and is the current editor of the journal *Human Biology*.

ANTON B. FISTANI founded the Laboratory of Human Paleontology and Prehistory at the Universiteti Luigj Gurakuqi in Shkoder, Albania in 1982 and remains as its current Director. He took his doctorate in Biology and Chemistry, and since 1977 has dedicated himself to carrying out research in the area of human paleontology and prehistory. In 1981 he discovered and then began excavating the palaeolithic site of Gajtan that he describes in this paper, which is the first description of the site to be published in the West. He has explored other areas in Albania and has discovered the additional potential hominid sites of Baran, Bleran and Rragam and the quarry of

Shahinove. Fistani is the only Albanian at present who is carrying out explorations and research in human paleontology. In 1992 he received a Fullbright Fellowship to spend the year studying comparative vertebrate paleontology at the Balcones Research Center, University of Texas, Austin. Fistani has published a number of research papers on his research, but this is his first presentation in English to the West.

LASZLO KORDOS is a Paleontologist at the Hungarian Geological Institute in Budapest, Hungary. He has worked at the oldest paleoanthropological site in Hungary, Rudabanya, publishing studies on the paleoenvironment, the paleontology and the prehominid fossils *Rudapithecus hungaricus*. He has also published on the biostratrigraphy of the Hungarian hominid site, Vértesszölös. His more general research has been on Hungarian vertebrate biostratigraphy and other studies on vertebrate paleontology in Central Europe. Kordos is Executive Director in Hungary of the Satos Foundation which provides research funding to East and Central Europeans.

ALEXANDER KOZINTSEV is a Physical Anthropologist at the Institute of Ethnography, St. Petersburg, Russia. He has read and published abstracts in Russian on some 2,000 articles, reports and books that have been published in Physical Anthropology in English and other languages, thus making available to his colleagues a great deal of research that might otherwise have been passed over. His research is in the area of skeletal biology, specializing in the analysis of cranioscopic traits. He is interested in devising new approaches for the analysis of nonmetric variation, and has applied his methods to studying skeletal data from many populations such as Eskoaleut, Chukchi, and investigating the affinities of the Mongoloid populations of Siberia, and of population mixture in Japan. In broader perspective, he is interested in studying the population history of Siberia and the Far East. He has published numerous articles in these areas; in many of these he has presented overviews of the research areas which are extremely insightful and useful summaries for the reader. He has the creative knack of being able to look at a problem and to present it to an audience in an understandable, well thought out manner which grasps the heart of the problem and lays it gently before the reader.

VALENTYNA LEONOVIČOVÁ, born in the Soviet Union, now a citizen of Czechoslovakia, received her D.Sc. at the State University of Leningrad in the Psychology Faculty. Her thesis topic was "Man as a Subject of Scientific Research," and she has continued this interest in her subsequent research. She taught in Siberia, then Leningrad, before becoming a Research Associate at the Laboratory of Evolutionary Biology of the Czechoslovak Academy of Sciences in Prague, Czechoslovakia in 1975. She has many publications which focus on the role of social factors in human evolution and the relationship between the biological nature of Man and his social environment. After moving to Czechoslovakia, her main area of research became the question of behaviour as a form of adaptation, and its role in evolution and sociobiology. She has been involved in a number of international symposia in the field of the Science of Man, throughout Europe, both as organizer and participant.

JANUSCZ PIONTEK is a Physical Anthropologist at the Institute of Anthropology, University of Poznań, Department of Human Evolutionary Biology, in Poznań, Poland. His research is in human evolutionary biology and his interests include biocultural evolution, the notion of "norm" in dealing with problems in modern human biology, the history of Anthropology in Poland, and methodological approaches to the study of prehistory and human evolution, to mention only a few areas. He is editor of a book recently published entitled *The Peculiarity of Man*, which presents current views on sociobiological and biocultural interpretations on the uniqueness of Man.

BECKY A. SIGMON is Professor of Anthropology at the University of Toronto, Canada. She is a paleoanthropologist whose research in human evolution has focussed primarily on issues relating to the origin and evolution of erect bipedal posture in humans. Since 1984 she has been interested in the question of socialism and its effects on the development of science. Two symposia were organized to explore this question, the first in Czechoslovakia in 1989, and the second in Toronto in 1991 which led to the publication of the papers in this volume.

JAROSLAV SLÍPKA is a Professor in the Medical Faculty, Institute of Histology and Embryology at Charles University in Plzeń, Czechoslovakia. He has been teaching and carrying out research in anatomy for about 40 years. He has lectured and given papers at conferences throughout Europe, and is currently President of the Anatomical Society of Czechoslovakia. His most recent research has involved the accumulation of a vast amount of information on human birth defects which he has tied in with the increase in environmental pollution. Other research includes a range of studies in embryology and anatomy which are concerned with the study of human evolution in general. Slípka has lived through 4 different political regimes, and throughout this time, has continued to teach and to carry out research with phenomenal optimism.

REM I. SUKERNIK is the Head of the Laboratory of Human Population Genetics in the Siberian Branch of the Russian Academy of Sciences, in Novosibirsk, Russia. Since 1971 he has participated in over 30 field expeditions to various parts of Northern and Southern Siberia, the Russian Far East and Middle Asia where he established field laboratories, compiled pedigrees and gathered a variety of medical and biological information on peoples of these areas. His laboratory research has been associated with nuclear and mitochondrial genetics, tissue culture, immunoserological and electrophoretic systems, blood typing, serum proteins, red call enzymes, immunoglobulin allotypes and other population polymorphic gene marker systems. Sukernik has earned an international reputation for his work in these areas. His publications are numerous and they have contributed valuable information to the field of human population genetics.

HERBERT ULLRICH, Physical Anthropologist, is presently at the Institut für Anthropologie der Humboldt-Universitat in Berlin, Germany where he was relocated in 1992. Before this move, and while still in East Germany, he managed to carry out a remarkable amount of research in human evolution. He has written extensively in the field of Paleoanthropology, and was the primary person responsible for organizing the interdisciplinary working group on "Probleme der Menschwerdung" which was founded at his institute, the Zentralinstitut für Alte Geschichte und Archaologie der Akademic der Wissenschafter der DDR, in East Berlin, 1977. This group, which he headed, included over 30 specialists from various disciplines, all of whom were devoted to working on problems on the origin and evolution of humans and human society. In his current position in the united Germany, he is initiating the Human Evolution International Interdisciplinary Project "Man and the Environment in the Palaeolithic" whose major aim is "...to provide a broad interdisciplinary dialogue and a deeper interdisciplinary cooperation in studying human evolution on an international level."

R. Sukernik   H. Ullrich

A. Fistani   J. Radovčić   L. Kordos

J. Slípka   V. Leonovičová

A. Kozintsev   M. Butovskaya   V. Leonovičová

  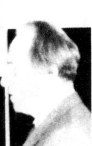

H. Helmuth   J. Radovčić   H. Ullrich

L. Kordos, E. Szathmary, J. Piontek (in foreground)

Back Row from left to right:

R. Sukernik
L. Kordos
H. Ullrich
J. Slípka
V. Leonovičová
M. Butovskaya
A. Kozintsev
A. Fistani

Front Row from left to right:

B. Sigmon
J. Piontek
J. Radovčić

PREFACE

The origin of this symposium goes back to 1984 when I made my first trip behind the "Iron Curtain," to Czechoslovakia to attend a conference on "Evolution and Morphogenesis." My primary purpose in going to this conference was to hear papers on a subject that was not a popular topic of research in North America, epigenetics. One of the major themes at the Czech conference was "Epigenetics," and C.H. Waddington who was a prime mover in this field (Waddington, 1957), was being honoured in memorium. Invited to speak on the topic was Søren Løvtrup, a Swedish embryologist, who had written many papers on epigenetics as well as a book on the topic (1974). Although quite well known in Europe for his research, Løvtrup has received less notice in North America.

The Czech conference was organized by one of our Toronto participants, Jaroslav Slípka. It was my first meeting with this multi-talented and highly capable individual, but it was to be followed by a number of subsequent meetings, I am happy to say. This was also my first meeting with Valentyna Leonovičová, the co-organizer of the present Symposium. We discovered then and at subsequent meetings, that we share a number of interests in human evolutionary studies, in spite of having extremely different backgrounds.

At the conference in Czechoslovakia in 1984, I couldn't help noticing differences in its organization and academic presentations in comparison to "Western" meetings. These differences seemed to me sufficiently important to explore further, and I made a mental note to do so when the opportunity presented itself.

Subsequent conferences that I attended on topics of human evolutionary research, were in the Soviet Union and East Germany in 1987. It was during this combined trip that I met Alexander Kozintsev at the Institute of Ethnography in Leningrad where I went to look at the Kiik-Koba skeletal remains. This trip was also the occasion that I met Herbert Ullrich in Halle, East Germany, at a symposium on *Homo erectus* where research papers were given on the nearby site of Bilzingsleben that was being excavated by Ursula and Dietrich Mania (Mania, 1981; 1987). Again, it seemed striking to me that the approaches of the socialist countries differed significantly from that of the "West." These differences will be described in my paper in this volume.

In 1989 the opportunity arose to co-organize a conference in

Czechoslovakia, with Leonovičová. Another conference was being held there in August of that year (the 3rd Anthropological Congress of Ales Hrdlička), and we decided to encourage people who were attending the Hrdlička Congress to participate in ours, that we entitled "Different Approaches to the Study of Human Evolution: East and West." The fact that the meeting was being held in Czechoslovakia meant that scientists from Socialist countries could attend without going through the problems of obtaining Western currency, visas and permission from proper authorities, which were sufficient reasons to discourage attendance. Thirty-five scholars from 15 countries attended, presented papers, and exchanged ideas and information of the work they were carrying out in their respective countries (see Appendix at the end of this volume for list of participants and titles of presentations). Although some of these papers are being or have been published in the journal *Human Evolution*, primarily the symposium served as a preliminary meeting for establishing rapport between colleagues from capitalist and socialist countries interested in the Science of Man. Everyone at the conference agreed that it would be a worthwhile goal to continue the theme of this conference, and to hold a second symposium at a location in North America. The main purpose would be to acquaint the West with research in Socialist Europe, and to do so in English which has become the international language of science. Native English speakers are generally less likely to read publications in East European languages, than the reverse is the case. Bringing East and Central European scientists together in North America where they present information in English about their research, was considered a positive step in furthering communication between East and West.

The 1989 meeting in Czechoslovakia occurred in September. Our timing was both fortuitous and imminent as a forecast for the immediate future, for in one month the Berlin Wall fell. One by one, Socialist European countries began making a major shift in political structure, and the consequences of this are still being felt around the world. The year afterwards, as economy and life in these countries began to settle down, seemed an auspicious time to initiate the proposed second symposium.

I applied for and received two grants, from the Wenner-Gren Foundation for Anthropological Research and from the Natural Sciences and Engineering Research Council, to organize a symposium in Toronto entitled "Physical Anthropology in East and Central Europe: Past, Present and Future Directions." Ten participants from 7 countries were invited. Selection of participants was based on several factors. They represent scientists who are prominent and have made valuable contributions to research in their country in the field of Physical Anthropology, and they are people who have shown an active interest in the approach taken in the 1989 and present 1991 symposia themes. We received advice from North American and European colleagues as well as relied on our own experience and knowledge of researchers active in this field. We would like to have invited many others equally competent, but lack of funds made this impossible.

The Symposium on East and Central European Physical Anthropology was held in Toronto in October, 1991. The purpose of the symposium was twofold. First, it was to provide information about human biological research that has been carried out in Socialist and post-Socialist Europe. Second, it was to explore the questions of the effect of political ideology on the Science of Man, and the changes in this science that are occurring as the countries undergo changes in political structure. In this context, subsidiary questions arise such as what kind of influence will increased contact with the West have on these countries, and conversely, what will be the effect on the West with increased communication with post-Socialist Europe?

I began organizing the symposium in the spring of 1991, and was particularly alert when I heard the BBC news broadcast one June morning. Gorbachev was reported to have stated that, although the Soviet Union must, and is, undergoing changes, it will not become simply a copy of the West. In this same vein, Václav Havel, the first elected President of Czechoslovakia after the "Gentle Revolution" of November, 1989, has also expressed concern that Czechoslovakia should not switch over to unbridled capitalism as it is expressed in the West (Havel, 1992). Both leaders seem to feel that a variation on the West's approach could be found for their countries.

These comments brought me back to the present moment, as I reflected on what will be the effect, especially in science, of changes in the Soviet Union, Czechoslovakia and the other European countries that have experimented with Socialism. The present is an exciting and exceptional time in the world of science. So much that was closed is now open. The past is still so much a part of the present that it is essential to have an understanding of it in order to realize the potential of the future. I hope that this Symposium will be one of the stepping stones toward achieving this end.

Becky A. Sigmon
Toronto, November, 1992

REFERENCES

Havel, V. 1992. *Summer Meditations*. Alfred A. Knopf, Canada.

Løvtrup, S. 1974. *Epigenetics—A Treatise on Theoretical Biology*. John Wiley, London, UK.

Mania, D. 1981. *Homo erectus* in middle Europe: the discovery from Bilzingsleben. In B. Sigmon and J. Cybulski, eds. *Homo erectus Papers in Honor of Davidson Black*. Univ. of Toronto Press, Toronto, pp. 133-151.

————. 1987. Homo erectus from Bilzingsleben (GDR): His Culture and his Environment, *Anthropologie* 25(1):1-45.

Waddington, C.H. 1957. *The Strategy of the Genes*. J.W. Arrowsmith Ltd., Bristol, Great Britain.

CHAPTER 1

THE EFFECTS OF SOCIALISM ON THE SCIENCE OF MAN*

B.A. SIGMON

INTRODUCTION

For some time now I have reflected on the tendency of North Americans, indeed perhaps "Westerners" in general, toward a Western focus in science. Although ethnocentrism is supposed not to be a characteristic of science, all scientists work within a cultural framework and their approaches are affected by the milieu in which the research is carried out. I first became aware of this tendency in myself after several years of teaching courses in Anthropology that dealt with the topic of human evolution. As my classes became more and more a mixture of racial types (Toronto being a major centre that attracts immigrants from all parts of the world), it occurred to me that the evolutionary theory and concepts in human biology that I was teaching and applying to my research, represented thought that came from only one part of a varied world. Evolutionary theory had developed in England and Western Europe, spread to North America, and grew with these flavours of influence.

As my classes came to have more and more first and second generation East and Central European students as well as Africans and Asians, whose parents or who themselves had been born and reared in a different political and/or cultural environment from the West, I paused to reflect on how the evolution of humans was considered in their place of origin. It especially

---

* I have decided to use the word "Man" in this book because it is the term more commonly used in East and Central European works in this field. In translation, English has no equivalent general term for Man (people or humans come closest but are not quite equivalent), as do the East and Central European languages where the specific term for man, the male, differs from the general term, Man.

bothered me that there was nothing oriental in the evolutionary theories that I taught. Occidentals had developed an explanation for evolution that dealt with physical evolution. The theories were based on being able to study processes scientifically, that is, through analysis of biology and through a process of replication of any given analysis. If an experiment could not be replicated, it was considered not scientific. This "Western" approach to studying natural history, termed the scientific approach, necessarily imposed limits on the kind of natural phenomena that could be studied. Traditionally, the Orient did not limit their approach to understanding life in this way; they were more willing to consider events that could not be exactly replicated. Their approach to explaining life was not limited just to the "scientific method" (e.g., see Chang, 1986).

In the 20th century, with the spread of a communist approach within the politico-economic system of East and Central European countries, one of the results was the closing of the borders of those countries. "Iron curtain" was the Western term used to describe this phenomenon. Effectively, this meant that academic communication was restricted or even forbidden by socialist political leaders, a decision that is deadly for the scientific approach which thrives on the open sharing of discoveries, data and ideas. As a consequence of border closing and restriction in communication, studies of science in capitalist and socialist countries began developing more or less independently of each other. By analogy to Oriental and Occidental countries, the capitalist and the socialist countries operated to some degree as two distinctly closed systems. However, this situation was different from the Oriental and the Occidental systems in that "science" was already known and had been practised in both capitalist and socialist countries before the latter had changed their political structure. Inevitably, the way that science continued to be developed and research conducted in both systems, began to differ. The socialist countries operated in relative isolation and behind closed walls. Indirectly this resulted in the setting up of barriers to the West, as they kept their scientific results mainly within the confines of these imposed barriers (thus the term "iron curtain" through which one can look vaguely but penetrate only with difficulty).

How science developed differently in these two systems is the focus of this paper. Since I am addressing primarily a North American audience, and I myself am a product of this system, this paper will concentrate not on the development of the science of Man in the West (we all know about that), but rather will focus on how the study of human evolutionary research, and indeed the Science of Man, developed in the now ex-socialist countries of Eastern and Central Europe.

One of the questions that I, as well as others, have asked is how has socialism affected the development of science in general, and specifically how has it affected studies in the human biological sciences? Since socialism makes certain assumptions about what is best for human beings (as do all ideologies — it is just the assumptions that differ), it is of interest to us here to see how the science dealing with human biology has been treated by the government, and how scientists themselves have responded.

There is no ultimate or simple answer to this inquiry, as one might have anticipated. When I discussed this question with my colleagues from different socialist countries, I concluded that part of the answer is found through looking at the past history of each country. The traditions that were already present, and the year that the country became socialistic, affected the way that socialism was applied there, both by the government and by the scientist. This observation may seem to be perfectly obvious in hindsight, but arriving at such a conclusion has taken quite a lot of reflection and time spent listening to my colleagues.

The papers in this volume are written by scientists in human biological studies from six European countries including the former USSR (Soviet Union), CSSR (Czechoslovakia), DDR (East Germany), Hungary, Poland and Albania. A presentation on "Paleoanthropology in Yugoslavia" was given at the Toronto Symposium by Jacob Radovčić, but unfortunately we have not received his manuscript to be able to include it here, although I will make reference to his presentation in this paper. Also, where relevant, I will discuss the content of certain presentations that were given at the 1989 symposium in Czechoslovakia (also organized by Leonovičová and myself – see my preface in this volume). The contents of all of these papers are, in a sense, statements that address the above question of how socialism has affected the Science of Man.

SCIENCE OF MAN BEFORE THE WALL FELL

Under socialism, research in the Science of Man continued, but its development was affected by the political ideology as interpreted by each country that enveloped socialism into its political structure. Since my own experience has been primarily with the Soviet Union and Czechoslovakia, my paper will draw more heavily on examples and events of these two countries. In addition, I will try to summarize the general development of human biology of the other countries represented by participants of this symposium.

THE SOVIET UNION

The name Soviet Union will be used here since the papers presented were written while the Soviet Union still existed intact; the USSR continued to exist until the early days of 1992 when the Union broke apart and the largest country remaining was Russia which is the place of origin of the authors in this book.

SOVIET GENETICS

Perhaps the most devastating academic effect of the Socialist experiment was that which occurred in Soviet genetics between 1934-1964 (for greater detail, see Medvedev, 1969). The views of Lysenko, an agricultural biologist, began to take precedence over other thought in genetics, including human genetics. Lysenko's view which stated that new forms could be "created" within the short time of a few generations, was appealing to the Stalinist government because it seemed to be the means that would provide quick, easy and better agricultural returns for the country. In addition, it provided a way by which a new type of person could be "created" and fit into the Communist mould. This invalid view of genetics ignored structural aspects of inheritance such as genes and chromosomes, as well as Mendel's laws, and relied on the theory that traits that were acquired in one generation could be passed on to succeeding generations.

The Soviet Academy of Sciences was forced into endorsing the Lysenko approach. This led to the demise and ultimately the destruction of Soviet genetics, which had been very advanced in its researches up to that time. This ideological approach also led to the dismissal and even arrest of renowned creative thinkers such as Vavilov who was arrested and sent to prison in 1940 (Medvedev, 1969). N.I. Vavilov, whose thinking was extremely advanced for genetics at the time, is referred to in this volume by Kozintsev as the Founder of Modern Biology and Genetics in the Soviet Union.

Another great thinker of his time, I.I. Schmal'gauzen (also written Schmalhausen — in Russian the h is pronounced as a g), whose impressive book on evolution, *Faktorii i Evolutii*, was published in Russian in 1947, was arrested because the book was regarded as going against official political doctrine. During his trial in 1948, his book was being translated into English, and it was smuggled out of the country and published under the title *Factors in Evolution* in the West in 1949. Schmal'gauzen lost his professorship, and his books and research projects were ordered destroyed. Claiming that he was not a geneticist, he was allowed to take a lesser position as senior research worker in the Zoological Institute of the Academy of Sciences working in a politically non-threatening area of comparative embryology and anatomy of fishes and amphibians (Schmalhausen, 1986).

Dogmatic positions change with time, as can be seen in the case of Schmal'gauzen, for at the Czech conference in 1984 on "Evolution and Morphogenesis" (Mlikovsky and Novak, 1985), several papers were given on the significance of the role that Schmal'gauzen's work had in evolutionary theory (Vorob'eva, 1985; Smirnov, 1985), indicating that, with his political acceptance, there had begun to be a renewal of interest and pride in this Russian scientist who had been demoted during the Lysenko period.

In this volume, Sukernik and Crawford refer to the devastating effects that the Lysenko approach had on the science of genetics, to say nothing of the disastrous results that it had on agriculture when the theory did not work in the growing of crops. Mendelian genetics was allowed back into the curriculum of the university in the latter part of the 1950s, and became an acceptable idea in research after that time. However, human genetics was not to recover until very recently (see Sukernik and Crawford, this volume) because party doctrine continued to believe that anything that fell under the rubric of human nature — such as human genetics — should be controlled by upper eschelons of political structure.

Leonovičová (this volume) confirms this dogmatic approach which was translated into official doctrine. In her paper, she reiterates the fact that any academic discipline that included studies associated with human nature was suspect. Anyone who supported the belief that there could be a biological basis for behaviour was, as she describes, guilty of biologism (*biologizatorstvo*), or following the ideology of the "enemy." The Science of Man was censored by ideological thought; anything having to do with the nature of Man was to be handled by political authorities, not academics.

SOVIET PRIMATOLOGY

Primatological studies in the USSR were not immune from the ideological stances taken by the Communist Party. Both the direction of research and the conclusions that were made by scientists were strongly influenced by party politics. According to Butovskaya (this volume), the period most strongly affected by party politics occurred between 1948-1964, although from the late 1930s onward, as in genetics, party doctrine was to influence the field of Soviet Primatology. The early work on primates, and Primatology has a long history in the Soviet Union, had begun with freedom from ideological intervention. Psychologists like Ladygina-Kots began behavioural studies on chimpanzees in 1923, and her work was followed by the establishment of a primate research facility in 1927 at Suchumi, Georgia, for purposes of studying primate behaviour and also for investigating medical primatology. This centre was the first in the world directed toward the study of medical primatology. Human evolutionary research was also a focus during this early period, as Soviet scientists were among the earliest scientists to recognize that studying our primate relatives can provide us with insights into our own evolution and processes of thinking.

However, the latter was the undoing and the termination of this early period of primate research. By the 1940s any research that had to do with human processes of thinking, and human nature, was considered top priority research that was to be controlled by ideological branches of the Party. Control of the human psyche to encourage the development of the collective "nature" of humans, rather than individual incentive, was of major importance to the ideology of Soviet Communism. The success of this, even after 70 years, can be read in Smith's *The New Russians* (1991) in chapters dealing with economic change. The idea of collective responsibility, ingrained in Russian thinking by 70 years of Communist doctrine, Smith shows has developed to such an extent that those who show individuality and incentive are spurned and criticized, their efforts blocked, or they may be ostracized from the group. This effectively squelches the success of *perestroika*, or economic reform.

The ideology that promoted this concept of collective thought led to the rejection of the theories of Mendelian genetics and Darwinian (and Schmal'gausian) evolution. These theories were replaced by Engel's labour theory of human origins, Lysenko's inheritance of acquired traits, and Pavlov's conditioned reflexes (see Butovskaya, this volume).

This period during which ideological dogma dominated academic life affected the research that was carried out, and the conclusions that scientists made. Butovskaya points out how the direction of thought of Ladygina-Kots, for example, changed under the influence of this dogma (as she compares the 1959 research with that from 1923). The pressure on scientists to support the official party doctrine greatly impeded the development of primatology during this time.

In the 1960s, although there was a thaw in this dogmatic official doctrine when theories of Mendel and Darwin were once again viewed as acceptable, the field of primatology which had nearly collapsed, was slow in rising again. There were no courses of instruction or training programs available at universities. And although new studies were initiated for studying the physical and behavioural adaptive abilities of apes, they developed in virtual isolation from those in the West. This was to be the condition of Soviet Primatology throughout the 1980s as well. Although lectures on primatology were resumed at Moscow State University in 1978, and new projects were initiated at the Suchumi Primate Centre, these studies continued to be carried out in relative isolation from research going on in other parts of the world.

The disadvantages that Communist dogma inflicted on Soviet Primatology that impeded its development from the late 1930s until recently are obvious in this discussion. However, it would be easy to let it go at this, and not see any positive effects of the imposed dogma and isolationism. As Soviet Primatologists enter into communication with the West, we may also see that they have developed points of view and insights into primate biology and behaviour that will be different from studies in the West. Comparisons of approaches, and what can be learned by each from the other may prove to be as useful to the West as it is to Soviet primatologists.

SOVIET PHYSICAL ANTHROPOLOGY

Kozintsev (this volume) gives us a slightly more encouraging view of the development of Physical Anthropology in the Soviet Union. Studies on the biological aspects of humans continued to be carried out for the duration of the existence of the Soviet Union. However, there are some interesting points that he has made that might be pursued further, to give us insight into the effect that party dogma had on the development of this discipline and the broader subject of Anthropology in general.

The study of Physical Anthropology, like that of human genetics and primatology, was affected in certain ways, during the dogmatic stage of Socialism from the late 1930s to the late 1950s and into the 1960s. First, publication of the *Journal of Anthropology* was terminated. Secondly, V.V. Bunak who is considered the Father of Physical Anthropology in the Soviet Union, or who Kozintsev describes as "everyone's teacher," and who also founded the Institute of Anthropology in 1923, was forced out of it in 1948. This was as significant a loss in Physical Anthropology as was the termination of Vavilov in Genetics.

Another area that was affected by party dogma was the direction in which Physical Anthropology grew. Certain research subjects including those having to do with the biological bases of human nature, thought processes and cognitive abilities, were not allowed. The fields that did develop had to do more with population concepts (thus fitting better with the dogma of the collective rather than the individual characteristics of human nature). Research was carried out on human population genetics, palaeoanthropology (at a more descriptive level, but not infringing into areas of individual incentive or development — for example, see Alexeev, chapters 6 and 7, 1986), growth and constitution, morphology, physiology, racial variation and human adaptability, the origin of language and other biological population studies.

In the Soviet Union, Anthropology existed only in the form of Physical Anthropology (Kosintsev, 1992). That to say, the general subject of Anthropology in the Soviet Union was considered to be what Physical Anthropologists study, not what North Americans have come to think of as Anthropology, which includes a broader spectrum of fields including Linguistics, Archeology and Social/Cultural Anthropology. In other Western countries, Physical Anthropology is often placed in the Faculty of Medicine or Anatomy, whereas Anthropology per se is considered to be primarily Social and Cultural Anthropology. From a North American point of view, other fields of Anthropology that developed in the Soviet Union, simply grew up under different names such as Ethnology and Archeology.

In summary, communist dogma affected the development of the Science of Man in the USSR, in the field of genetics, primatology, psychology and physical anthropology. This began in the mid-1930s and continued for varying times in the different academic disciplines. It devastated the study of genetics, restricted the type of primatological studies that were carried out, and influenced the direction of the study of Physical Anthropology.

CZECHOSLOVAKIA

In Czechoslovakia, the effect of communist dogma applied to academics was more strongly felt after 1968 than before, because a more liberal form of socialism was in practice from 1948-68 (Wheaton and Kavana, 1992).

During the two-year period of 1968-69, when Czechoslovakia was forced to become more rigid in its application of socialism, many intellectuals left the country. Those who stayed behind were to be affected, some more and some less, by official doctrine that also moved into academic halls.

Although I cannot include all those scientists who continued in the study of the Science of Man, I would like to elaborate on the responses of a few people, and how they continued to carry out their research under the strict socialist regime, especially as it was initiated after 1968. Since Paleoanthropology is discussed in this paper in the next section, I will limit the discussion here to those scientists who specialized in human evolutionary topics other than paleoanthropology.

One very influential scholar was V.J.A. Novak, an entymologist internationally known for his work in insect physiology. Novak established, in 1976, the Department of Evolutionary Biology in the Czechoslovak Academy of Science. Its purpose was to study origins of life and general questions of evolution. After 1978 its focus enlarged to include the general topics of ontogenetic development, the role of environment and behaviour in evolution and primatology (Vančata, 1991). This attracted students of human evolution and of anthropology in general. Novak was able to add to his department: the Russian sociobiologist V. Leonovičová and molecular biologist Z. Masinovsky, and the Czech Anthropologists/Human evolutionists V. Vančata, V. Novotny and K. Zemek.

Czechoslovakia, as Slipka discusses in his paper in this volume, has a long intellectual tradition. Its survival, in the middle of Europe and with no natural physical barriers to keep out invaders, has depended in part on an aptness of mind and flexibility to accept change without in fact its peoples being changed. Irony has been a secret "weapon," one might say, that the Czechs have developed to an art. (One of the best examples of this is Hašek's *The Good Solidier: Schweik*). Their emphasis on the importance of polyglotism epitomizes this attitude which has ensured the survival of the identity and integrity of the people. An unstated tradition is that if one understands another's language, then one understands better the people who speak that language. Understanding, then, has been used as a means of adaptation by the Czechs.

With this as his cultural heritage, Novak took the initiative and made opportunities for the development of his Department of Evolutionary Biology in socialist Europe. He was even able to reach into the West and touch some scientists who shared interests in evolution with him (e.g., F.J. Ayala, S.W. Fox, M.W. Ho, H.J. Jerison, S. Løvtrup, M. Ruse, P. Saunders and myself). He did this partly by organizing international conferences on major themes in evolutionary theory. Between 1978 and 1987 he was one of the major thrusts behind the organization of six international conferences on evolution. (Novak, Lenovičová and Pacltova, 1979; Novak and Mlikovsky, 1982; Novak and Zemek, 1983; Mlikovsky and Novak, 1985; Novak, Vančata and Vančatova, 1986; Leonovičová and Novak, 1987), thus making his Department well known and respected especially in socialist Europe. Jaroslav Slípka was involved with the 1984 conference, and undertook the majority of the organization. The lives of these two scientists have been linked for many years.

Both Novak and Slípka are persons who have lived through four changes of government in Czechoslovakia during their lifetimes — and each are people of great integrity who have contributed to the Science of Man in their own ways. Novak was a dedicated Communist believing in its idealism. Slipka remained a non-Communist, quietly practising his own philosophy, and contributing in his own personally adapted ways. To understand how two colleagues can differ in political ideology, but can remain friends, mutually respecting each other, as Novak and Slípka did, one might read Havel's article concerned with the question "Can a man be honest, courageous, true to his conscience and still be a communist?" (Havel, 1991; also see Havel, 1992).

Novak, as head of the Department of Evolutionary Biology, and as a dedicated Communist, had an effect on students and on those who worked in his department. On the one hand, he was respected and admired. His direction toward internationalizing the Department was a good move for all. His emphasis to his students and co-workers to learn language, travel abroad (with restrictions) and do their own work was encouraging and has prepared a generation of colleagues to begin participating in the scientific world outside of socialist Europe (e.g., Mlikovsky, Novotny, Vančata , Vančatova and Zemek). On the other hand, there was a negative side as well, as there usually is with any strong leader. Although he encouraged original evolutionary thinking, he also advocated his own views of human evolution which were strongly Marxist in content (Novak, 1982 and Leonovičová, this volume). The younger scientists, and those others who were working under him, could not always agree with Novak's views, yet they had to be cautious in their criticisms. This period made it difficult for them, caused conflict which resulted in perhaps stopping their publications or travel abroad, and indeed to some extent affected their own freedom of expression in making interpretations in their own work.

There are probably always disadvantages associated with advantages. Especially when a person is a strong leader and innovative worker and thinker, there will be conflicting opinions about his behaviour and his character and how much influence he should have in his discipline. One could compare the influence of Novak with that of S.L. Washburn, the American Physical Anthropologist, who trained a generation of Physical Anthropologists in the 1950s, 60s and into the 70s. Both were influential in their countries and at overlapping times.

Washburn encouraged his students to pursue newly developing areas of Physical Anthropology such as primate behaviour and studies of anatomy and morphology as a means of learning more about human evolution. Some of his students have been, e.g., Tuttle, Zihlman, Jay, Lancaster, all major workers in the field of Physical Anthropology in the United States. Washburn's first position was at the University of Chicago, following which he moved to the University of California at Berkeley where he worked until he retired. He was a person who could foresee the need for developing new areas of human evolutionary research, and he was an apt synthesizer. Washburn is probably best known through his students and because he was a major visionary in Physical Anthropology. Although not always right in his conclusions, and not always liked for his dogmatic positions, he was the person who coined the term, the "new" Physical Anthropology in 1951 (Lasker, 1965), and encouraged its development. Both scientists, Novak in Czechoslovakia and Washburn in the United States, played a major role in the direction of research for the succeeding generation (Washburn's influence was mainly in the 1950s to the 70s, and Novak's, from the 1960s to the 80s).

In making concluding remarks about the effect of socialism on human biological studies in Czechoslovakia, I wish to turn to comments made by Slípka, both in his paper in this volume and remarks that he made to me personally. It is evident from Slípka's review that the intellectual tradition of studies in the human biological sciences continued in spite of changes in the government to socialism. Scientists in his country quietly pursued studies in anatomy, embryology and morphology, and in some respects they have become leaders in this area, by North American as well as by European standards. As he himself says, the study of morphology in Czechoslovakia "...means a science of developmental causes of structures and forms, i.e., of their evolutionary history."

The continuing development of the study of human morphogenesis occurred partly because of the socialist policies of isolationism and stricter control of academic curriculum. One of the results of this was limited accessibility of technologically advanced equipment such as the West had begun using; such equipment was just not as readily available to Czech scientists. Free exchange of scientific information between the West and socialist Europe was also restricted. The focus of research that had been carried out both pre- and post-World War II, continued to be a major thrust in the following decades. This fact has often been interpreted by Western scientists as meaning that scientific development in socialist Europe is "behind" the West by several decades. This, however, disregards the continued study and advances in thinking that have been made in a field that the West nearly abandoned because of their turning increasingly to the use of sophisticated equipment for interpreting scientific data generated by other sophisticated equipment. Studies of human morphogenesis and embryology were on the demise in the West at the time that sophisticated studies and research in countries like Czechoslovakia continued to be carried out in this field.

The Czech pride in their accomplishments, in spite of the restrictions and isolation imposed by their "totalitarian regime" is evident in Slípka's personal remark to me (Slípka, 1991). We had been comparing the student instruction in anatomy and embryology at the university level, and he was convinced that theirs was the more instructive education. Very generously, he suggested to me that we send our students to Charles University in Plzen, so that they could thoroughly learn their anatomy, embryology and studies in morphogenesis. I was inclined to agree with him.

PALEOANTHROPOLOGY IN SOCIALIST EUROPE

The study of human origins seems to be an area that has such universal human appeal that it has remained a source of pride for countries to publicize information about fossil finds that have been discovered in their countries. No wall has been effectively placed around this area of science although there have been restrictions in East Germany and Albania. Fossils of our early ancestors, no matter where they are found, seem to be regarded as the heritage of all peoples, and in this sense, the field of paleoanthropology is an international endeavour. Of course, since the study of fossils, to a great extent, deals with interpretations on morphology and phylogeny, this discipline did not directly affect or threaten political doctrine. It is only in the area where behavioural interpretations of the fossils are attempted, that some suppression might be experienced or where ideology might affect the behavioural interpretation (e.g., see Alexeev, 1986; Hermann and Ullrich, 1991 and Ullrich, this volume).

Czechoslovakia, Hungary and Yugoslavia have long traditions in paleoanthropological research. J. Jelínek and E. Vlček of Czechoslovakia are two primary figures who have worked a lifetime in paleoanthropology. Aside from carrying out a great deal of research in fossil Man, Jelínek established and has continued duties as editor of *Anthropologia: an International Journal of the Science of Man* in which have been published a number of research articles from socialist Europe. Vlček has been involved in a number of works dealing with fossil hominids (e.g., Vlček, 1969; 1979; 1981; Mania and Vlček, 1981). Kordos (in this volume) has reviewed paleoanthropological research in Hungary, and has shown its long tradition in the country (also see Eiben, 1988). Radovčić (1988a; 1988b) of Yugoslavia has continued the tradition of fossil hominid research at the Krapina site that was discovered in 1899 by Gorjanovic-Kramberger. The work done by Kramberger on Krapina fossil hominids was a landmark in human paleontology because it was he who first began applying "...standard, yet at that point still unusual, comparative and descriptive methods, implementing a number of technological and scientific innovations in analyzing paleontological material" (Radovčić, 1988a: p.12).

In East Germany, in spite of little government support and even with restrictions imposed on some work, research was carried out in human origins with quiet determination (considering the conditions) by scientists like Ullrich and Mania. At the same time, little research in this area was being done in West Germany. In 1977 Ullrich organized a group of 30 scientists from East Germany from a variety of disciplines, to make up what he termed "Menschwerdung." Their purpose was to study the origins and early adaptations of humans. This group is still in existence in the united Germany, although since 1990 people outside of the country are also included, and Ullrich has co-authored a book on the topic which is considered to be an extensive coverage of the field (Hermann and Ullrich, 1991). Again, the effect (now become stigma) of ideology surfaced when the book was criticized for the excess of labour views of evolution, (the book was published in the united Germany) which some suggested were due to the authors' heritage from communist doctrine in East Germany.

Also in East Germany, Dietrich Mania worked on excavating a *Homo erectus* site near a village called Bilzingsleben (from 1976 to the present). In spite of meagre funding and little government support, he and Ursula Mania have continued to discover very useful information about the hominids that lived during the time period of 350,000 years ago. And they managed to get the information published (e.g., see Mania, 1980; Mania and Vlček, 1987; Mai et al., 1983; Mania and Weber, 1986). Mania himself was regarded as a "dissenter" who, as a result, was denied permission to travel out of the country. In spite of the imposed isolationist policy that personally affected him, he continued to carry on his work at Bilzingsleben. Scientists from outside of East Germany came to visit his site and discover for themselves what results he had found, and in this way, the information was disseminated outside.

ALBANIA

Albania represents an entirely different situation from the rest of socialist Europe. It is a country bordered by other regions where fossils have been found, and its own location is a prime area for the past habitation of hominids although few fossils have been found there yet. Even before communism was brought into Albania in 1944, which made it virtually isolated from the West, neither Paleoanthropology nor Physical Anthropology were traditions there. Much of the research in prehistory and general topics on human evolution had been carried out by foreign scientists such as Gobineau (1853), Haberland and Lebzeller (1919), Pittard (1921), Tildesley (1934) and Weininger (1934). It wasn't until the Congress of Illyrian Studies in 1972 in Tirana that the Albanian Academy of Science acknowledged the urgent need to begin studies in the anthropological sciences. Primarily these were regarded as studies into the Illyrian origins of

Albanians and Albanian prehistory. Even the prehistory, as Fistani (1992) points out, was limited mainly to the study of stone tools as archeologists were trained mostly as historians rather than natural scientists. Physical remains of Pleistocene Man, faunal data, information on sedimentology and palynology were generally neglected "because these researches were so expensive and were considered a kind of pure science without any utilitarian result" (Fistani, 1992).

After the Congress of Illyrian Studies in 1972, several researchers began investigations into the Illyrian origins of Albanians through taking measurements made on living Albanians and comparing them with skeletal populations from Albanian archeological sites (Ylli, 1975; Dhima, 1979; 1982; Cipi, 1984). This research, referred to as Ethnic Anthropology, differed from that which Anton Fistani (Fistani, 1991) began carrying out. In 1982, Fistani established the Laboratory of Human Paleontology and Prehistory at the University of Luigi Gurakugi in the northern town of Schkoder. Fistani had already spent several years working at a late Pleistocene site, Gajtan, near Shkoder, that he had discovered. I first learned about Fistani's research through the efforts of the Candian film producer, Paul Jay who had spent 4 years working on a documentary film on Albania (Jay, 1991). Jay brought manuscripts, which were summaries of Dhima's and Fistani's work, out of the country and passed them on to me. According to my knowledge, it was the first time that these research findings had been received in North America. Aside from this, the little bit of information that was known about Albanian anthropological research had been received through letters that had been passed on to foreign travellers who mailed them from outside of Albania.

Because of his research in human origins, we invited Fistani to present his findings at the Symposium (see his paper, this volume). It was the first time in the West that he had spoken about his research. Previous efforts of his to leave the country for conferences or research purposes had been unsuccessful. His research was not considered by the state to be of high priority. In spite of the isolationism imposed on him, he carried out this work over a period of 15 years. In addition, he has explored sites in other parts of the country and presents a good case for the possibility of their being additional early hominid sites in Albania. Exploration of these in the near future, should prove to be extremely important to palaeoanthropology. At the present time, 1992, research has been put on hold as the country goes through the difficult period of making the transition from socialism. Fistani himself (in 1992) is in the United States on a Fullbright Fellowship.

HUNGARY AND POLAND

Since I have not visited Hungary or Poland, I must rely primarily on the reports of my colleagues, with special reference to the papers prepared by Kordos and Piontek in this volume. In both countries, there was a strong emphasis on the development of Human Biological research; in the West, this is a branch of Physical Anthropology that concentrates primarily on the biology of living peoples. Studies in prehistory have also been carried out, but a great deal of emphasis has focussed on human biology. Eiben has summarized Hungarian anthropology and human biological research (1982; 1988), and Bielicki et al. (1985) have prepared a summary of the same for Poland. In this volume, Piontek reviews the vast amount of research in Poland that has been devoted to the biology of living peoples. He states that the fields of Social and Cultural Anthropology are only of "marginal" interest to Polish Anthropologists. A possible interpretation for this may be that socialist ideology discouraged investigations into human behaviour and human nature (as it did in the Soviet Union), which is an area of research that more often falls under Social and Cultural Anthropology. Physical Anthropology, especially with an emphasis on paleoanthropology and human biology, would perhaps have been a less controversial area of research, and especially the latter would have contributed useful information on people that could be applied by the state in the making of policies.

In summary, both Hungarian and Polish scientists continued to carry out research in the Study of Man during the period that their countries were under socialism. Their primary limitations were isolation from the West, and pressure from the Communist Party to conform to what it thought was best for academics (Eiben, 1989).

CONCLUSION

With these examples, I have tried to provide some insight into the question initially asked on how political ideology can affect science. The society that we live in does affect our choices of study and research, and influences (but does not limit) what we consider important. Some subjects of study are more and others less influenced by ideology. Those areas tending to be under greater control are related to human behaviour and economic output.

In the field of Physical Anthropology, the study of Paleoanthropology seems to have been "claimed" as an internationally neutral discipline, belonging to all countries and all peoples. No matter what the official political doctrine, the fossil study of early humans has been and continues to be of international interest, and political barriers have not placed a wall around the information relating to these fossil finds. This has been true for Eastern and Central Europe, for Communist China, and always for the mother continent of human fossils—Africa and its countries.

The country most affected by communist ideology is the one in which it was first implemented, Russia. For a generation and more it lost some of the sciences that were developing in other countries, such as genetics, some aspects of primatology and physical anthropology.

Did political ideology affect the human spirit of inquiry? This is not an easy question because there is no simple or single answer. Ideology might influence the direction that an inquiry takes because one operates within the context of one's environment. But dogmatic ideologies cannot, and do not, control the spirit of inquiry which is and always remains, a basic feature of human nature. This we have seen through examples in this book. The creative thinkers who are scientifically minded and who ask questions about life and nature, cannot be prevented from seeking the truth, nor can they be prevented from spreading the word about discoveries that they make.

Because science is based on the human spirit of inquiry, to seek the hows and whys of nature, it is universal and transcends geographic and political boundaries. Seemingly, like the study of our earliest human fossil ancestors, it is something that belongs to the whole human race (see Bronowski's *Science and Human Values*, 1956). If East and Central Europe's experiment and their time with Socialism has taught us anything, it is that in spite of ideological practices that censor, restrict, abolish or impede the scientific search for knowledge, there will be people who will find a way to continue their research and they won't be still or content with less.

REFERENCES

Alexeev, V.P. 1986. *The Origin of the Human Races*. Progress Publishers, Moscow.

Bielicki, T., T. Krupinski and J. Strzalka. 1985. *History of Physical Anthropology in Poland*. International Assoc. of Human Biologists, Occas. Papers: 1 (no. 6).

Brownski, J. 1959. *Science and Human Values*. Harper and Row, New York.

Chang, Kwang-Chih. 1986. *The Archaeology of China*. 4th ed. Yale Univ. Press.

Dhima, A. 1979. *A propos du type anthropologique des albanais durant le Moyen Age*, pp. 301-142 Iliria IX-X, Tirane.

_____. 1982. *Te dhena antropologjike per gjenzen e shqiptarve*, Ph.D. dissertation.

Eiben, O.G. 1982. One hundred years of the Budapest Department of Anthropology. *Human Biol*. Budapest, 9, 17-29.

_____. 1987. Changing patterns of growth, development and aging in a Hungarian population. *Coll. Anthrop.* (Zagreb), 11, 73-90.

_____. 1988. *History of Human Biology in Hungary*. International Assoc. of Human Biologists, Occas. Papers: 2 (no. 4).

_____. 1989. Personal communication.

Fistani, A. 1992. Personal communication.

Gobineau, Comte de. 1853 *Essais sur l'inégalité des races humaines*. Quatrième ed. Librairie de Paris, t. II.

Haberland, A. and V. Lebzelter. 1919. *Zur Physischen Anthropologie der Albaner. Archiv fur Anthrop*. 123, XVII.

Hašek, Y. 1964. *The Good Soldier: Schweik*. F. Ungar Publ. Co., New York.

Havel, V. 1991. Thinking about Frantisek K. *The Idler*. No. 32: 20-22.

_____. 1992. *Summer Meditations*. Alfred A. Knopf, Canada.

Hermann, J. and H. Ullrich. 1991. *Menschwerdung: Millionen Jahre Menschheitentwicklung - natur - und geistes wissenschaftliche Ergebnisse*. Akademie Verlag, Berlin.

Jay, P., Director. 1991. *Albanian Journey: End of an Era*, High Road Production.

Lasker, G. 1965. The "new" physical anthropology seen in retrospect and prospect. *Centennial Rev. of Arts and Sciences*: 348-366.

Leonovičová, V. and V.J.A. Novak, eds. 1987. *Behaviour as One of the Main Factors of Evolution. Proc. International Symposium*, Czechoslovak

Academy of Sciences, Praha.

Løvtrup, S. 1974. *Epigenetics—A Treatise on Theoretical Biology.* John Wiley, London, UK.

Mai, D.H., C. Mania, T. Notzold, V. Toepfer, E. Vlček and W.D. Henrich. 1983. *Bilzingsleben II. Homo erectus—seine Kultur und seine Umwelt.* Veroff. Landesmus. Vorgesch. Halle, Vol. 36.

Mania, D. 1980. *Bilzingsleben I. Der geologische Bau des Travertinkomplexes auf der Steinrinne bei Bilzingsleben.* Veroff. Landesmus. Vorgesch. Halle, Vol. 36.

————. and T. Weber. 1986. *Bilzingsleben III. Homo erectus-seine Kultur und seine Umwelt.* Veroff. Landesmus. Vorgesch, Halle, Vol. 39.

————. and E. Vlček. 1981. *Homo erectus* in middle Europe: the discovery from Bilzingsleben. In B. Sigmon and J. Cybulski, eds. *Homo erectus: Papers in Honor of Davidson Black.* Univ. of Toronto Press, Toronto, pp. 133-151.

————. and E. Vlček. 1987. Homo erectus from Bilzingsleben (GDR)—His Culture and his Environment. *Anthropologie* 25 (1): 1-45.

Medvedev, Z.A. 1969. *The Rise and Fall of T.D. Lysenko.* Columbia University Press, New York.

Mlikovsky, J. and V.J.A. Novak. 1985. *Evolution and Morphogenesis. Proc. International Symposium,* Plzen, 1984.

Novak, V.J.A. 1982. *The Principle of Sociogenesis.* Praha: Academia, 214 pp.

————., V. Leonovičová and B. Pacltova, eds. 1979. *Natural Selection. Proc. International Symposium, Liblice, 1978.* Czechoslovak Acad. of Sciences, Praha.

————. and J. Mlikovsky. 1982. *Evolution and Environment. Proc. International Symposium, Brno, 1981.* Czechoslovak Academy of Sciences, Praha.

————. and K. Zemek, eds. 1983. *General Questions of Evolution. Proc. International Working Colloquium, Liblice, 1982.* Czechoslovak Academy of Sciences, Praha.

————., V. Vančata and M. Vančatova, eds. 1986. *Behaviour, Adaptation and Evolution.* Czechoslovak Academy of Sciences, Praha.

Puttard, E. 1921. *Découverte de l'age de la pierre en Albanie. Recherche anthropologie dans la peninsula des Balkans, specialement dans la Dobroudja.* Vol. I.

Radovčić, J. 1988a. *Dragutin Gorjanovic—Kramberger and Krapina Early Man: the Foundations of Modern Paleoanthropology.* Skolska Knija, Zagreb.

————. 1988b. *The Krapina Hominids: An Illustrated Catalog of Skeletal Collection.* Croatian Natural History Museum, Zagreb.

Schmal'gauzen, I.I. 1949. *Factors of Evolution.* The Blakiston Co., Philadelphia.

————. 1947. *Factorii i Evolutii.* Muskva. Ivd. AN SSSR, 396 pp.

Schmalhausen, I.I.. 1986. *Factors of Evolution.* Univ. of Chicago Press, Chicago.

Slípka, J. 1991. Personal communication.

Smirnov, I.N. 1985. Ivan Schmal'gauzen and the development of the evolutionary doctrine. *Evolution and Morphogenesis. Proc. International Symposium*, Plzen. Ed. by J. Mlikovsky and V. Novak. pp. 187-194.

Smith, H. 1991. *The New Russians*. Random House, New York.

Tildesley, M. 1934. The Albanians of the North and South. *L'Anthropologie*, I (XLIV). Abstract published in *Biometrika* 25: 21-55.

Vančata, V. 1991. Personal communication.

Vlček, E. 1969 *Neadertaler der Tschechoslowakei*. Praha.

———. 1979. *Homo erectus bilzingslebensis*. Eine neue Form des mittelpleistozanen Menschen in Europa. *Ethnogr. Archaol. Z.* 20: 634-661.

———. 1981. Fossiler Mensch aus Weimar-Ehringsdorf. Vortragsmanuskript Internationale Tagung. *Anthroposoziogenese*, Weimar.

Vorob'eva. E.I. 1985. *Evolutionary morphology and the theory of evolution. Evolution and Morphogenesis. Proc. International Symposium*, Plzen 1984. Ed. by J. Mlikovsky and V.J.A. Novak. 171-180.

Waddington, C.H. 1957. *The Strategy of the Genes*. J.W. Arrowsmith Ltd., Bristol, Great Britain.

Weininger, J. 1934. Rassenkundliche Untersuchungen an Albanern, ein Beitrag zum Problem der dinariscshen Rasse. In R.P. Nachlass, serie A. *Physische Anthropologie*, Bd. IV, Anthropologische Gesellschaft, Wien. Vol. 4, pp. 1-68.

CHAPTER 2

CURRENT DEVELOPMENTS IN SOVIET PHYSICAL ANTHROPOLOGY

A.G. KOZINTSEV

ABSTRACT

Major directions and trends in Soviet physical anthropology are described. From the late 50s to the mid-70s, an important orientation shift has occurred, with increasing emphasis on population biology.

INTRODUCTION

There is little doubt that physical anthropologists in the USSR are better informed about the work conducted by their North American colleagues than the latter know about Soviet research. This is due primarily to isolationist tendencies which were imposed on Soviet science by the bureaucratic regime and prevented Soviet scholars from publishing their results in the West and even from supplying their articles and books with extensive English summaries — but partly also to a certain reluctance of many North American scientists to read anything not written in English. The analysis of reference lists shows this quite distinctly (Kunter, 1980).

The aim of the present paper is to give some information about the state of Soviet physical anthropology. My task was facilitated by a recent review by Turner (1987) who presented what might be termed a surface survey. As a complement, "a look from within" may also be useful.

THE FOUNDERS

Tracing back the professional genealogies of Soviet physical anthropologists, we ultimately reach a person who may with full right be regarded as "everyone's teacher" — V.V. Bunak (1891-1979). All scholars who have been and are engaged in this discipline in the USSR are his pupils, either directly or indirectly. The scope of his research was enormous, covering virtually every major area — paleoanthropology, population genetics, origin of speech and growth and constitution, to mention only a few. His last fundamental monograph was published after his death (Bunak, 1980).

The influence of G.F. Debetz (1905-1969) was also enormous. He concentrated on racial studies and achieved extremely much, having investigated dozens of populations all over the USSR as well as abroad, from the Caucasus to Chukotka and from North Siberia to Afghanistan. His professional reputation was hardly lower than that of Bunak (Debetz was Secretary General of the ICAES). Representing a different tradition, he was Bunak's long-standing opponent.

The tradition to which Debetz belonged was initiated by A.I. Yarkho, a brilliant theorist who died in 1935, at the age of 32. His ideas on racial analysis produced a profound effect upon an entire generation of Soviet anthropologists.

The fourth founder was Ya.Ya. Roginsky (1895-1986) whose principal area of research was early human evolution. His writings are not numerous but they are marked by exceptional insight. One of the best examples is provided by his last small essay on the origin of art (Roginsky, 1982).

Other outstanding representatives of the older generation were M.G. Levin, V.P. Yakimov, N.N. Cheboxarov, and M.M. Gerasimov.

THE TRANSITION

Important developments which resulted in the emergence of the present phase of Soviet physical anthropology had very little to do with the usual mechanism responsible for the change of paradigms, as described by T.S. Kuhn — older scientists leaving the stage to make room for the young. In fact, very few of our predecessors were burdened with archaic ideas (individual racial typology, for example, has never been popular in this country). Moreover, it was Bunak who did a great deal to link anthropological research with principles of population biology as early as the 1920s and 1930s.

However, the conditions were far from favourable. In 1937, *The Journal of Anthropology* (the only one!) was closed. The situation with genetics became worse and worse since the mid-30s and, in 1948, the final disaster broke out, with fatal consequences for all related disciplines. Bunak had to leave the Institute of Anthropology which he created in 1923. These events put an end to all attempts of modernizing physical anthropology.

A thaw which followed Stalin's death and the 20th congress of CPSU has affected all sciences. In 1957, a new journal emerged, initially titled *Soviet Anthropology* and later *Problems of Anthropology (Voprosy antropologii)*; it should be noted that, in U.S.S.R., the term *anthropology* means "physical anthropology." Until the present, it is the only journal in our country solely devoted to this discipline. Being published by the Moscow University, it is predominantly an organ of the Institute of Anthropology and the Chair of Anthropology (both belonging to the University). However, for most of its history, it was open to specialists representing all sub-disciplines, without regard to institutional affiliation, and only recently the situation has changed for the worse. So the contents of this journal may give some idea of what was happening in this area.

My analysis is based on 1229 articles published in 92 issues of *Voprosy antropologii* from 1957 to 1990. The topics of these articles are listed in Table 1, and the number of articles belonging to each topic is presented for each of eleven 3-year periods, with percentage freqeuncies in parentheses. Note that the topics are not mutually exclusive (one article may, and usually does, belong to two and more of them). Besides the 20 topics (a few minor ones are omitted), three additional characteristics, pertaining to number and sex of authors, are considered.

The frequencies, after being transformed into angles measured in radians, were subjected to principal components analysis. The first principal component (PC1) accounts for almost half (46.3%) of the total variance and thus represents the major tendency. Since this is the only component showing a meaningful diachronic trend (see below), other components will not be considered. Correlations of topics with PC1 and the scores of this component for each period are given in Table 1.

The characteristics with the highest negative loadings (Fig.1) are frequencies of articles devoted to history of anthropology, brain, non-human primates, comparative anatomy, and skeleton (especially fossils). The characteristics with highest positive loadings are frequencies of joint articles (especially those having four and more authors), and articles concerned with anthropometry, physiology, genetics, adaptation, pathology, growth, and body composition. This polarity reveals the contrast between traditional physical anthropology and a newer science which P.T. Baker (1982) termed human population biology (the emphasis on the population aspect is beyond doubt in our case since otherwise the articles would normally not be accepted for publication). The results also demonstrate that the new discipline, contrary to its traditional counterpart, is developed by research teams rather than by individual scientists. The share of women in human biological investigations is higher than it is in traditional physical anthropology (are they better team members?). Institutional affiliations were not analysed, but the situation appears clear at first sight: while the proportion of "outsiders," such as geneticists or medical practitioners, may be somewhat higher in human population biology, even in this area the most important research is done by anthropologists.

TABLE 1

REPRESENTATION OF TOPICS IN *VOPROSY ANTROPOLOGII*, 1957-90, AND
RESULTS OF THE PRINCIPAL COMPONENTS ANALYSIS

TOPICS	1957-59	1960-62	1963-65
Non-hominid primates	21 (24.7)	19 (14.3)	13 (12.3)
Comparative anatomy	15 (17.6)	11 (8.3)	2 (1.9)
Pre-Holocene human evolution	13 (15.3)	23 (17.3)	16 (15.1)
Post-Pleistocene geographic variation	28 (32.9)	42 (31.6)	24 (22.6)
Neurology	6 (7.1)	12 (9.0)	11 (10.4)
Osteology	27 (31.8)	49 (36.8)	27 (25.5)
Anthropometry and anthroposcopy of the living people	6 (7.1)	22 (16.5)	16 (15.1)
Constitution	4 (4.7)	5 (3.8)	2 (1.9)
Body composition	1 (1.2)	2 (1.5	9 (8.5)
Dental studies	3 (3.5)	3 (2.3)	2 (1.9)
Dermatoglyphics	5 (5.9)	4 (3.0)	1 (0.9)
Biochemical polymorphisms and population genetics	4 (4.7)	7 (5.3)	13 (12.3)
Genetics of somatic traits	0 (0.0)	0 (0.0)	1 (0.9)
Physiology	5 (5.9)	8 (6.0)	13 (12.3)
Growth and aging	6 (7.1)	16 (12.0)	20 (18.9)
Behaviour	8 (9.4)	12 (9.0)	11 (10.4)
Adaptation	0 (0.0)	1 (0.8)	3 (2.8)
Social and professional groups	1 (1.2)	2 (1.5)	2 (1.9)
Pathology	0 (0.0)	0 (0.0)	1 (0.9)
History of Anthropology	9 (10.6)	10 (7.5)	4 (3.8)
Joint articles (2 and more authors)	9 (10.6)	15 (11.3)	11 (10.4)
Joint articles (4 and more authors)	0 (0.0)	0 (0.0)	0 (0.0)
Women authors	40 (43.5)	68 (48.2)	55 (51.4)
Total no. of articles	85	133	106
Component score on PC1*	**-6.03**	**-5.03**	**-2.51**

*   Characteristics of 1229 articles published in 92 issues of *Voprosy Antropologii*, 1957-1990: correlations with the first principal component.

TABLE 1 (CONTINUED)

1966-68	1969-71	1972-74	1975-77	1978-80
20 (14.6)	11 (8.1)	21 (13.5)	14 (9.7)	21 (16.3)
3 (2.2)	5 (3.7)	7 (4.5)	6 (4.1)	7 (5.4)
12 (8.8)	16 (11.9)	19 (12.3)	9 (6.2)	18 (14.0)
23 (16.8)	40 (29.6)	43 (27.7)	49 (33.8)	42 (32.6)
6 (4.4)	4 (3.0)	1 (0.6)	4 (2.8)	10 (7.8)
33 (24.1)	23 (17.0)	34 (21.9)	20 (13.8)	25 (19.4)
21 (15.3)	40 (29.6)	50 (32.3)	51 (35.2)	42 (32.6)
5 (3.6)	12 (8.9)	12 (7.7)	6 (4.1)	11 (8.5)
7 (5.1)	9 (6.7)	9 (5.9)	15 (10.3)	7 (5.4)
6 (4.4)	3 (2.2)	6 (3.9)	9 (6.2)	8 (6.2)
9 (6.6)	15 (11.1)	14 (9.0)	18 (12.4)	13 (10.1)
15 (10.9)	21 (15.6)	25 (16.1)	22 (15.2)	10 (7.8)
1 (0.7)	4 (3.0)	4 (2.6)	9 (6.2)	4 (3.1)
23 (16.8)	26 (19.3)	40 (25.8)	32 (22.1)	15 (11.6)
26 (19.0)	20 (14.8)	37 (23.9)	25 (17.2)	19 (14.7)
10 (7.3)	11 (8.1)	8 (5.2)	8 (5.5)	15 (11.6)
8 (5.8)	9 (6.7)	13 (8.4)	10 (6.9)	8 (6.2)
4 (2.9)	7 (5.2)	7 (4.5)	2 (1.4)	2 (1.6)
4 (2.9)	3 (2.2)	11 (7.1)	11 (7.6)	6 (4.7)
10 (7.3)	5 (3.7)	8 (5.2)	4 (2.8)	3 (2.3)
22 (16.1)	34 (25.2)	52 (33.5)	50 (34.5)	48 (37.2)
2 (1.5)	6 (4.4)	14 (9.0)	11 (7.6)	8 (6.2)
67 (46.5)	69 (44.2)	101 (48.8)	107 (53.2)	90 (49.5)
137	135	155	145	129
-0.87	1.51	2.81	3.62	0.55

TABLE 1 (CONTINUED)

1981-83	1984-86	1987-90	TOTAL	CORRELATION WITH PC**
7 (10.1)	8 (13.8)	7 (9.1)	162 (13.2)	-0.6
5 (7.2)	2 (3.4)	4 (5.2)	67 (5.5)	-0.53
11 (15.9)	8 (13.8)	11 (14.3)	156 (12.7)	-0.49
29 (42.0)	16 (27.6)	34 (44.2)	370 (30.1)	0.17
3 (4.3)	3 (5.2)	2 (2.6)	62 (5.0)	-0.69
18 (26.1)	18 (31.0)	24 (31.2)	298 (24.2)	-0.61
19 (27.5)	15 (25.9)	24 (31.2)	306 (24.9)	0.88
5 (7.2)	8 (13.8)	6 (7.8)	76 (6.2)	0.55
6 (8.7)	10 (17.2)	11 (14.3)	86 (7.0)	0.73
3 (4.3)	2 (3.4)	6 (7.8)	51 (4.1)	0.42
3 (4.3)	6 (10.3)	4 (5.2)	92 (7.5)	0.62
15 (21.7)	7 (12.1)	7 (9.1)	146 (11.9)	0.80
2 (2.9)	2 (3.4)	0 (0.0)	27 (2.2)	0.83
15 (21.7)	10 (17.2)	8 (10.4)	195 (15.9)	0.88
14 (20.3)	11 (19.0)	16 (20.8)	210 (17.1)	0.75
4 (5.8)	8 (13.8)	6 (7.8)	101 (8.2)	-0.34
11 (15.9)	13 (22.4)	9 (11.7)	85 (6.9)	0.83
1 (1.4)	3 (5.2)	0 (0.0)	31 (2.5)	0.35
2 (2.9)	2 (3.4)	0 (0.0)	40 (3.3)	0.80
1 (1.4)	0 (0.0)	2 (2.6)	56 (4.6)	-0.76
27 (39.1)	23 (39.7)	25 (32.5)	316 (25.7)	0.89
7 (10.1)	5 (8.6)	2 (2.6)	55 (4.5)	0.92
66 (62.9)	47 (49.5)	63 (63.0)	773 (50.5)	0.42
69	58	77	1229	-
2.63	2.63	0.69	-	-

**   The first principal component based on 23 characteristics of articles published in *Voprosy Antropologii*, 1957-1990.

These results pertain mostly to the Moscow University where, since the times of Bunak, the biological emphasis in anthropology was very strong. At the second major centre, the Institute of Ethnography, which belongs to the Academy of Sciences and has two branches, in Moscow and Leningrad, the emphasis is on ethnical rather than biological problems, and the studies are focussed on racial origins and population history, following the tradition introduced by Debetz. The same holds for anthropological centres in other cities since most of them are headed by pupils or followers of Debetz. Scholars working at these centres were (and, theoretically, are) able to publish their results in *Voprosy antropologii*, yet more often they used other journals. No quantitative survey of such journals was attempted. There is no doubt, however, that over the last decades important changes have occurred in these as well. Specifically, there was a considerable growth of studies in population genetics, dental anthropology, and dermatoglyphics.

THE PRESENT STATE

The brief review that follows is based on publications which have appeared in 1980-90; some works published in 1991 could also be included. For lack of space, only books are mentioned. So our survey pertains to accomplished major projects (it should be noted that the lag between the completion of a study and its publication is 2-3 years at best in our country) rather than works in progress. Preparing a comprehensive account of the current research would demand too much time and effort.

Primatological studies focussed on the relevance of primate behaviour for early hominid evolution (Fainberg, 1980; Deryagina, 1986; Godina, 1991). Most of the observations and experiments were done at the Primate Centre in Sukhumi, Georgia, and used to reconstruct social relations in hominoid communities, morphofunctional preconditions for bipedal locomotion, and origins of tool use.

Studies on early humans are represented by two monographs, one describing the mandibular fragment of the Mindel-Riss age from Azykh, Azerbaijan (Kasimova, 1986), another, Upper Paleolithic skeletons from Sungir', Central Russia (Zubov and Kharitonov, 1984). Descriptions of other new Pleistocene fossils from the USSR are given in Lazukov (1981), Roginsky (1989), and Godina (1991). Specific as well as general issues of hominid taxonomy are discussed by Alexeev (1985b).

Morphological variation in modern man, mainly at the individual level, is reviewed by Nikityuk and Chtetsov (1991) with regard to all organs and systems. Similar reviews pertain to the skull (Speransky and Zaychenko, 1980; Speransky, 1988) and the soft tissues (Bekov, 1988). Khrisanfova (1990) has examined the relations between human physique and sex hormone levels (see also Mazhuga and Khrisanfova, 1980).

At the population level, extensive statistical analyses of anthropometric variation in adults were performed with special reference to sex differences

(Kurshakova et al., 1983). Deryabin and Purunjan (1990) continued this research, with main stress on the multivariate analysis of geographical patterns in body build. The first author published a textbook of multivariate methods for physical anthropologists (Deryabin, 1983).

Books dealing with population history and racial classification, based on studies of living people as well as skeletal series, are quite numerous (Alexeeva et al., 1986 catalogued the skeletal collections of the Moscow University). Some publications are primarily methodological (Gokhman, 1980a; Alexeev, 1985a; 1985b; 1989). Others present data pertaining to large parts of the world (Alexeev and Trubnikova 1984), the USSR (Alexeev and Gokhman, 1983; 1984; Gokhman, 1980b; 1984), and separate regions such as the European part of USSR (Zubov and Shlygina, 1982; Kruts, 1984; Denisova et al., 1985; Gokhman and Kozintsev, 1986; Gerasimova et al., 1987; Kajanoja and Zubov, 1986; Cesnys and Balciuniene, 1988; Yusupov, 1989; Denisova, 1990; Aksyanova, 1991), Caucasus (Abdushelishvili, 1980; 1982); Turkestan (Kiyatkina, 1987; Babakov, 1988; Alexeev et al., 1984; Kiyatkina, 1987; Babakov, 1988; Alexeev and Bubova, 1989; Litvinsky and Zhdanko, 1991), Siberia (Okladnikov and Alexeev, 1980; Alexeenko et al., 1982; Zolotareva, 1982; Alexeeva et al., 1986; Alexeeva and Uryson, 1984; Alexeev, 1986; 1988; Velikanova and Zolotareva, 1986; Davydova, 1989; Polos'mak et al., 1989; Its and Tomilov, 1990), India (Abdushelishvili and Malhotra, 1980; 1982; Arutyunov, 1990), and China (Cheboxarov, 1982). The work on facial reconstruction, initiated by Gerasimov, is continued by his pupils (Zubov, 1988). Cranial nonmetric traits, very efficient for racial analysis, are presently being introduced by Kozintsev (1988).

The most significant event in dental anthropology is a comprehensive world-wide survey written by Zubov and Khaldeeva (1989). At the regional level, two monographs have appeared, on Latvia (Gravere, 1987), and Kazakhstan (Ismagulov and Sikhimbayeva, 1989).

The development of dermatoglyphics was marked by two fundamental analytical summaries, covering the USSR (Heet 1983) and the entire world (Heet and Bolinova 1990). The genetics of dermatoglyphic traits was investigated by Guseva (1986), and their distribution in Byelorussia, by Tegako (1989).

Population genetics is dealt with in important works by Rychkov and his associates, mostly focussing on Siberian populations (Rychkov and Sheremetyeva, 1980; later results are published in numerous articles in *Voprosy antropologii*). A monograph by Spitsyn (1985) summarizes his long-standing research on population aspects of biochemical polymorphism. A regional study was done in Byelorussia (Mikulich, 1989). Population genetical, as well as dental and dermatoglyphic, data may also be found in several of the books on population history mentioned above.

In growth studies, very significant results have been obtained by Miklashevskaya et al. (1988) who analysed the influence of environment on the development of children. Other authors who investigated various aspects of this problem are Dorozhnova (1983), Heapost (1984), Salivon et al. (1989). Environmental effects on the aging process in adults, with special regard to bone mineralization, have been studied by Pavlovsky (1987).

Research on human ecology and biological adaptation has been very extensive. Its results are summarized in several monographs, the most important of which is that by Alexeeva (1986), integrating the vast information obtained by her team in tracing the effects of various environments on human biology (see also Vasilevsky et al., 1980; Kaznacheev, 1983; Kaznacheev and Kaznacheev, 1986). Other publications focus on longevity (Bruk, 1982; Kozlov, 1987; 1989), pathology and selection (Altukhov, 1985; Alexeeva, 1989; Umnova, 1989), adaptation to physical exercise (Martirosov, 1982; Nikityuk and Kogan, 1989), and urban ecology (Alexeeva, 1990). Regional ecological studies have been conducted in Byelorussia (Tegako et al., 1981; 1982), Azerbaijan (Kozlov and Dubova, 1990), and Tuva (Alexeeva and Uryson, 1984).

THE PROSPECTS

The review presented above demonstrates that while traditional areas continue to flourish, the diversification process is probably going on, with newer and newer problems (and subsequently sub-disciplines) branching off the old stem of physical anthropology. In this respect, the situation in the USSR does not seem to substantially differ from that in the West. The question arises, whether it is still possible (or necessary) to consider physical anthropology a single science (which should now include human population biology) or whether the time has come to regard it as a confederacy of smaller sciences, such as ethnic anthropology, human population genetics, growth studies, anthropology of social and professional groups, medical anthropology, etc.; note, however, that, contrary to the Western tradition, all these sub-disciplines are strictly biological. The "lumping" tendency is advocated by Zubov (1982), the "splitting" tendency, by Nikityuk (see, e.g., Nikityuk and Tegako, 1983). Curiously, the debate arose simultaneously with, and independently from, the publication of an article by Baker (1982) declaring human population biology a transdisciplinary science.

Is the new science viable in the U.S.S.R.? Is its aggregation with traditional physical anthropology viable? The human factor appears to play a crucial role here. As long as the major part of research continues to be done by physical anthropologists having a common basic education and regarding themselves as direct or indirect pupils of Bunak, Debetz, Yarkho, and Roginsky, integrative tendencies will prevail, as demonstrated, for instance, by conferences in memory of Bunak (Alexeev and Zubov, 1986). But if too many "outsiders" with entirely different educational and professional backgrounds are absorbed (which does not seem to be the case at present), the risk of disintegration is likely to increase.

REFERENCES*

Abdushelishvili, M.G. 1980. *Anthropology of the Causcasus in the Feudal Period.* Metsniereba, Tbilisi. [Georgian]
_____. 1982. *Anthropology of the Caucasus in the Bronze Age.* Metsniereba, Tbilisi. [Georgian]
_____. and K.C. Malhotra (eds.). 1980. *New Data on the Anthropology of North India.* Nauka, Moscow.
_____. and K.C. Malhotra (eds.). 1982. *New Materials on the Anthropology of West India.* Nauka, Moscow.
Aksyanova, G.A. (ed.). 1991. *Origin of the Saams (Lapps): Anthropological and Archaeological Evidence.* Nauka, Moscow.
Alexeenko, E.A., I.I. Gokman, V.Vs. Ivanov, and V.N. Toporov (eds.). 1982. *Studia Ketica: Anthropology, Ethnography, Mythology, Linguistics.* Nauka, Leningrad.
Alexeev, V.P. 1985a. *Geographical Centres of Human Race Formation.* Mysl', Moscow.
_____. 1985b. *Man: Evolution and Taxonomy.* Nauka, Moscow.
_____. (ed.) 1986. *Problems of Anthropology of the Prehistoric and Recent Population of Soviet Asia.* Nauka, Novosibirsk.
_____. (ed.) 1988. *Paleoanthropology and Archaeology of West and South Siberia.* Nauka, Novosibirsk.
_____. 1989. *Historical Anthropology and Ehnogenesis.* Nauka, Moscow.
_____., T.I. Alexeeva, S.A. Arutyunov, and I.S. Gurvich (eds.). 1983. *At the Border Between Chukotka and Alaska.* Nauka, Moscow.
_____. and N.A Bubova (eds.). 1989. *Turkmens in the Turkestanian Interfluve.* Ylym, Ashkhabad.
_____. and I.I. Gokhman 1983. *Physical Anthropology of Soviet Asia.* In: Schwidetzky I. (ed.). Rassengeschichte der Menschheit 9. Oldenbourg, München.
_____. and I.I. Gokhman. 1984. *Anthropology of the Asiatic part of the USSR* Nauka, Moscow (revised and enlarged Russian version of the preceding monograph).
_____., T.K. Khodjayov, and Kh. Khalilov. 1984. *The Population of Upper Amu-Darya: Paleoanthropological Data.* FAN, Tashkent.
_____. and O.B. Trubnikova. 1984. *Some Problems of Taxonomy and Genealogy of the Asian Mongoloids: Craniometry.* Nauka, Novosibirsk.
_____. and A.A. Zubov (eds.). 1986 *Problems of Evolutionary Morphology*

of Man and his Races. Nauka, Moscow.

————. 1986. *Adaptive processes in human populations*. MGU, Moscow.

Alexeeva, T.I. (ed.). 1989. *Anthropology: Contributions to Medicine*. MGU, Moscow.

————. (ed.). 1990. *Urban Ecology*. Nauka, Moscow.

————., S.G. Efimova, and R.B. Ehrenburg. 1986. *Catalogue of Cranial and Osteological Collections of the Institute and Museum of Anthropology, Moscow State University*. (1st ed., 1979). MGU, Moscow.

————. and M.I. Uryson. 1984. *Anthropo-ecological Studies in Tuva*. Nauka, Moscow.

Altukhov, Yu.P. (ed.). 1985. *Human Physiological Genetics and the Problem of Wound Healing*. Nauka, Moscow.

Arutyunov, S.A. (ed.). 1990. *Origins of the Recent Population of South Asia*. Nauka, Moscow.

Babakov, O. 1988. *The Medieval Population of Turkmenistan*. Ylym, Ashkhabad.

Baker, P.T. 1982. Human Population Biology: A Viable Transdiciplinary Science. *Hum. Biol.* 54: 203-220.

Bekov, D.B. (ed.). 1988. *Individual Anatomical Variation of Organs, Systems, and Body Shape in Man*. Zdorovya, Kiev.

Bruk, S.I. (ed.). 1982. *Phenomenon of longevity*. Nauka, Moscow.

Bunak, V.V. 1980. *Genus Homo: Its Origin and Subsequent Evolution*. Nauka, Moscow.

Cesnys, G. and I. Balciuniene. 1988. *Anthropology of the Ancient Inhabitants of Lithuania*. Mokslas, Vilnius. [Lithuanian]

Cheboxarov, N.N. 1982. *Ethnic anthropology of China*. Nauka, Moscow.

Davydova, G.M. 1989. *Anthropology of the Mansi*. Nauka, Moscow.

Denisova, R.Ya. (ed.). 1990. *Balts, Slavs, Baltic Finns: Ethnogentical processes*. Zinatne, Riga.

————., J. Graudonis, and R.U. Gravere. 1985. *Kivutkalns Bronze Age cemetery*. Ainatne, Riga.

Deryabin, V.Ye. 1983. *Multivariate Biometrics for the Anthropologists*. MGU, Moscow.

————. and A.L. Purunjan. 1990. *Geographical Variations of Body Build in the Population of USSR* MGU, Moscow.

Deryagina, M.A. 1986. *Manipulatory Activity of the Primates*. Nauka, Moscow.

Dorozhnova, K.P. 1983. *Effects of Social and Biological Factors upon the Development of Children*. Meditsina, Moscow.

Fainberg, L.A. 1980. *The Origins of Social Relations*. Nauka, Moscow.

Gerasimova, M.M., N.M. Rud, and L.T. Yablonsky. 1987. *Anthropology of the Eastern European Populations in the Classic and Medieval Periods*. Nauka, Moscow.

Godina, Ye.Z. (ed.). 1991. *Primate Behaviour and Problems of Early Human evolution.* MGU, Moscow.

Gokhman, I.I. (ed.). 1980a. *Current Problems and New Methods in Anthropology.* Nauka, Leningrad.

————. (ed.). 1980b. *Studies in Paleoanthropology and Craniology of the USSR.* Collected Papers of the Museum of Anthropology and Ethnography 36. Nauka, Leningrad.

————. (ed.). 1984. *Problems of Anthropology of the Prehistoric and Recent Population of North Eurasia.* Nauka, Leningrad.

————. and A.G. Kozintsev 1986. *Anthropology of the Recent and Prehistoric Population in the European Part of USSR.* Nauka, Leningrad.

Gravere, R.U. 1987. *Dental Anthroplogy of the Letts.* Zinatne, Riga.

Guseva, I.S. 1986. *Morphogenesis and Genetics of the Papillary Skin in Man.* Belarus', Minsk.

Heapost, L. 1984. *Age Anthropology of Pupils at Tallinn Schools.* Valgus, Tallinn. [Estonian]

Heet, H.L. 1983. *Dermatoglyphics of the Peoples of USSR.* Nauka, Moscow.

————. and N.A. Bolinova. 1990. *Racial Differentiation of Mankind: Dermatoglyphical Data.* Nauka, Moscow.

Ismagulov, O. 1982. *Ethnic Anthropology of Kazakhstan (Somatological Study).* Nauka, Alma-Ata.

————. and K.B. Sikhimbayeva. 1989. *Dental Anthropology of Kazakhstan.* Nauka, Moscow.

Its, R.F. and N.A. Tomilov (eds.). 1990. *Anthropology and Historical Ethnography of Siberia.* Omsk University, Omsk.

Kajanova, P. and A.A. Zuboc (eds.). 1986. *Somatology and Population Genetics of the Bashkirs.* Annales Academiae Scientiarum Fennicae, Series A, V. Medica, No. 175. Suomalainen Tiedeakatemia, Helsinki.

Kasimova, R.M. 1986. *The Oldest Prehistoric Human Fossil Discovered in USSR: Azykh, Azerbaijan.* Elm, Baku.

Kaznacheev, V.P. 1983. *Essays on the Theory and Practice of Human Ecology.* Nauka, Novosibirsk.

————. and S.V. Kaznacheev. 1986. *Human Adaptation and Constitution.* Nauka, Novosibirsk.

Khodjayov, T.K. 1980. *Paleoanthropology of Uzbekistan.* Fan, Tashkent.

————. 1987. *Ethnic Processes in Medieval Turkestan.* Fan, Tashkent.

Khrisanfova, E.N. 1990. *Human Constitution and Biochemical Individuality.* MGU, Moscow.

Kiyatkina, T.P. 1987. *Paleoanthropology of Western Central Asia in the Bronze Age.* Donish, Dushanbe.

Kozintsev, A.G. 1988. *Ethnic Cranioscopy: Racial Variation of Cranial Sutures in Modern Man.* Nauka, Leningrad.

Kozlov, V.I. (ed.). 1987. *Abkhasian Longevity.* Nauka, Moscow.

————. (ed.). 1989 *Longevity in Azerbaijan.* Nauka, Moscow.

————. and N.A. Dubova (eds.). 1990. *Russian Old-timers in Azerbaijan, 1-*

2. Nauka, Moscow.

Kruts, S.I. 1984. *Paleoanthropological Studies in the Steppe Belt of Dnepr Valley.* Naukova Dumka, Kiev.

Kunter, K. 1980. Untersuchungen über die Literaturverzeichnisse anthropologischer Arbeiten, II. *Homo* 80: 212-232.

Kurshakova, Yu.S. (ed.). 1982. *Variation of Morphological and Physiological Traits in Males and Females.* MGU, Moscow.

————., T.N. Dunayevskaya, T.F. Durygina, A.L. Purunjan, T.P. Shagurina, S. Comenda, A. Martines, M. Rivero De La Calle (eds.). 1983. *Anthropometric Standardization of Populations in the SEV Countries.* Legkaya Promyshlennost', Moscow.

Lazukov, G.I. (ed.). 1981. *Nature and Prehistoric Man.* Moscow.

Litvinsky, B.A. and T.A. Zhdanko (eds.). 1991. *Problems of Ethnic Origins and Ethnic History of the Peoples of Turkestan, 4: Anthropology.* Nauka, Moscow.

Martirosov, E.G. 1982. *Methods of Investigation in the Anthropometry of Athletes.* Fizkul'tura i sport, Moscow.

Mazhuga, P.M. and E.N. Khrisanfova. 1980. *Problems of Human Biology.* Naukova Dumka, Kiev.

Miklashevskaya, N.N., Ye.Z. Godina, and V.S. Solovyova. 1988. *Growth Processes in Infants and Adolescents.* MGU, Moscow.

Mikulich, A.I. 1989. *Genetic Geography of the Rural Population of Byelorussia.* Nauka i tekhnika, Minsk.

Nikityuk, B.A. and V.P. Chtetsov (eds.). 1991. *Human Morphology* (First ed., 1983). MGU, Moscow.

————. and B.I. Kogan. 1989. *Skeletal Adaptation in Athletes.* Zdorovya, Kiev.

————. and L.I. Tegako (eds.). 1983. *Problems of Modern Anthroplogy.* nauka i tekhnika, Minsk.

Okladnikov, A.P. and V.P. Alexeev (eds.). 1980. *Paleoanthropology of Siberia.* Nauka, Novosibirsk.

Pavlovsky, O.M. 1987. *Biological Age of Man.* MGU, Moscow.

Polos'mak, N.V., T.A. Chikisheva, and T.S. Baluyeva. 1989. *Neolithic Cemeteries of Northern Baraba.* Nauka, Novosibirsk.

Roginsky, Ya.Ya. 1982. *On the Origins of Art.* MGU, Moscow.

————. (ed.). 1989. *Biological Evolution and Man.* MGU, Moscow.

Rychkov, Yu.G. and V.A. Sheremetyeva. 1980. The Genetics of Circumpolar Populations of Eurasia Related to the Problem of Human Adaptation. In: F.A. Milan (ed.). *The human Biology of Circumpolar Populations.* Cambridge University Press, Cambridge.

Salivon, I.I, N.I. Polina, and O.V. Marfina. 1989. *Infant Organism and Environment.* Nauka i tekhnika, Minsk.

Speransky, V.S. 1988. *Foundations of Medical Craniology.* Meditsina, Moscow.

————. and A.I. Zaychenko. 1980. *Shape and Construction of the Skull.*

Meditsina, Moscow.

Spitsyn, V.A. 1985. *Biochemical Polymorphism in Man: Anthroplogical Aspects.* MGU, Moscow.

Tegako, L.I. 1989. *Dermatoglyphics of the Byelorussian Population.* Nauka i tekhnika, Minsk.

————. and I.I. Salivon. 1982. *Ecological Aspects of Anthropological Studies in Byelorussia.* Nauka i tekhnika, Minsk.

————., I.I. Salivon, and A.I. Mikulich. 1981. *Biological and Social Aspects of Anthropological Variation.* Nauka i tekhnika, Minsk.

Turner, C.G. 1987. Physical Anthropology in the USSR Today, 1-2. *Quart. J. Archaeol.* June: 11-14; Fall: 4-6.

Umnova, M.A. (ed.). 1989. *Human Blood Group Systems and Transfusional Complications.* Meditsina, Moscow.

Vasilevsky, N.N., V.P. Kaznacheev, M.A. Kurdina, M.M. Mirrakhimov, N.P. Neverova, and A.D. Slonim (eds.). 1980. *Human Ecological Physiology.* Nauka, Leningrad.

Velikanova, M.S. and I.M. Zolotareva (eds.). 1986. *Ethnical Links Between the Populations of North Asia and America: Anthropological Data.* Nauka, Moscow.

Yusupov, R.M. 1989. *Craniology of Bashkirs.* Nauka, Leningrad.

Zolotareva, I.M. 1982. Anthroplogy of the Small Nations of Northern Siberia. *Annales Academiae Scientiarum Fennicae,* Ser. A, V. Medica. No. 174. Suomalainen Tiedeakatemia, Helsinki.

Zubov, A.A. 1982. The Contents of the Concept *Anthropology* on the Modern Level of Development and Integration of Science in the USSR. *Soviet Ethnography* 5: 21-33.

————. (ed.). 1988. *Physical Types of the Prehistoric Populations in USSR, Based on Facial Reconstructions.* Nauka, Moscow.

————. and N.I. Khaldeeva. 1989. *Dental Studies in Modern Anthropology.* Nauka, Moscow.

————. and V.M. Kharitonov (eds.). 1984. *Sungir': Anthropological Study.* Nauka, Moscow.

————. and N.V. Shlygina (eds.). 1982. *Studia Fenno-Ugrica: Anthropology, Archaeology, Linguistics.* Nauka, Moscow.

---

[*]   All references, if not otherwise indicated, are in Russian.

PHYSICAL ANTHROPOLOGY IN THE USSR: MAJOR RESEARCH CENTRES

In the Soviet Union, some research centres in physical anthropology are part of the university system belonging to the USSR Ministry of Higher Education while others are attached to the Academy of Sciences of the USSR The same structure exists at the level of separate republics (where it is becoming more and more independent from central stuctures). People working in Moscow and Leningrad normally do research in more than one geographical region, while those working in other cities focus on their own territories.

Moscow

Institute of Anthropology (103009 Moscow, pr. Marxa, 18). Director, V.P. Chtetsov (who is also head of the Chair of Anthropology at Moscow University and editor-in-chief of *Voprosy antropologii*), deputy director, V.P. Volkov-Dubrovin. This institute belongs to the Moscow University and is the only institute in the USSR specializing entirely in physical anthropology. The main emphasis is on human biology, and the principal areas of research are as follows:

Human ecology, climatic adaptation, physiology, body composition: T.I. Alexeeva, V.P. Volkov-Dubrovin, O.M. Pavlovsky, N.S. Smirnova, N.I. Klevtsova.

Biochemical polymorphisms and population genetics: V.A. Spitsyn, I.V. Perevozchikov.

Growth: N.N. Miklashevskaya, Ye.Z. Godina.

Anthropometry, standardization, biometry: Yu.S. Kurshakova, V.Ye. Deryabin, A.L. Purunjan.

Body build: V.P. Chtetsov.

Dermatoglyphics: T.D. Gladkova.

Osteology and racial history: T.I. Alexeeva, T.S. Konduktorova, S.G. Efimova (the last person is in charge of osteological collections).

Racial mixture: I.V. Perevozchikov.

Early human evolution: M.I. Uryson, V.M. Kharitonov.

Primate behaviour: V.I. Chernyshov.

MOSCOW UNIVERSITY BIOLOGICAL FACULTY, CHAIR OF ANTHROPOLOGY

(117234 Moscow, Leninskiye Gory, MGU).

Head, V.P. Chtetsov (see above).

The principal researchers are Ye.N. Khrisanfova (endocrinal aspects of body build; evolution of the human postcranial skeleton) and V.Z. Yurovskaya (comparative anatomy of soft tissues in the primates).

INSTITUTE OF ETHNOLOGY AND ANTHROPOLOGY, DEPARTMENT OF ANTHROPOLOGY

(117334 Moscow, Leninsky prosp., 32-a). This institute (formerly Institute of Ethnography) belongs to the Academy of Sciences. Contrary to the Institute of Anthropology, the research is directed mainly toward ethnic problems (racial history). The department is headed by A.A. Zubov. Following sub-disciplines are represented:

Dental anthropology: A.A. Zubov, N.I. Khaldeeva, G.A. Aksyanova.

Dermatoglyphics: H.L. Heet, N.A. Dolinova.

Anthropometry: I.M. Zolotareva, G.M. Davydova, G.A. Aksyanova, A.P. Pestryakov, N.A. Dubova.

Osteology: M.M. Gerasimova, G.V. Rykushina.

Facial reconstruction: G.V. Lebedinskaya, T.S. Baluyeva.

Population genetics: S.V. Sokolovsky.

Primatology: M. Butovskaya

INSTITUTE OF ARCHAEOLOGY

(117036 Moscow, ul. Dmitriya Ulyanova, 19). This institute also belongs to the Academy and is directed by V.P. Alexeev, the most productive among the living physical anthropologists in the USSR, author of several hundred works on early human evolution and racial history. However, the laboratory of physical anthropology at this institute is quite small. One of its members is L.T. Yablonsky who studies prehistoric populations of Turkestan.

INSTITUTE OF GENERAL GENETICS

(117809 Moscow, ul. Gubkina, 3). Here, a laboratory is headed by Yu.G. Rychkov who is the leading specialist in human population genetics in the USSR and has done extensive work on the genetics of Siberian populations.

INSTITUTE OF EVOLUTIONARY MORPHOLOGY AND ECOLOGY OF ANIMALS, LABORATORY OF GENETICS

(117071 Moscow): A.F. Nazarova (population genetics of Siberia).

INSTITUTE OF PHYSICAL CULTURE, CHAIR OF ANATOMY

(105483 Moscow, Sirenevy Bulvar, 4): B.A. Nikityuk (functional anatomy and genetics of somatic traits).

INSTITUTE OF LEGAL MEDICINE

(123242 Moscow, Sadovo-Kudrinskaya, 3, korp.2): V.N. Zvyagin (the leading expert in forensic osteology).

OTHER CITIES OF RUSSIA

LENINGRAD
INSTITUTE OF ETHNOGRAPHY,
DEPARTMENT OF ANTHROPOLOGY

(199034 Leningrad-34, Universitetskaya nab., 3). Formally, this is a division of the Moscow Institute of Ethnology and Anthropology, but in practice the Leningrad Department of Anthropology, headed by I.I. Gokhman, is independent from its counterpart in Moscow. The research, however, is also focussed on racial history, the following areas being represented:

Osteology: I.I. Gokhman, A.G. Kozintsev (who is in charge of skeletal collections), Yu.D. Benevolenskaya, A.V. Shevchenko, V.I. Khartanovich, Yu.K. Chistov, A.B. Radzyun.

Biochemical polymorphisms: Yu.D. Benevolenskaya.

Anthropometry, population genetics, dermatoglphics: V.I. Bogdanova.

UFA
INSTITUTE OF HISTORY, LANGUAGE AND LITERATURE

(450054 Ufa, pr. Oktyabrya, 71): R.M. Yusupov (osteology of the Bashkirs).

TOMSK
UNIVERSITY MUSEUM OF ARCHAEOLOGY AND ETHNOGRAPHY

(624010, Tomsk, pr. Lenina, 36): V.A. Dremov (osteology of West Siberian populations).

NOVOSIBIRSK
INSTITUTE OF MEDICAL PROBLEMS OF THE NORTH AND INSTITUTE OF CYTOLOGY AND GENETICS

(630090 Novosibirsk): R.I. Sukernik, T.M. Karaphet, S.V. Lemza, L.P. Osipova, O.L. Posukh and others investigate blood polymorphisms in the populations of Siberia.

INSTITUTE OF HISTORY, PHILOLOGY, AND PHILOSOPHY

(630090 Novosibirsk, pr. Lavrentyeva, 17): T.A. Chikisheva (osteology of Siberian populations).

OTHER REPUBLICS

TALLINN
ACADEMY OF SCIENCES OF ESTONIA,
INSTITUTE OF HISTORY
(200001 Tallinn, bulv. Estonia, 7):   L. Heapost (biochemical plymorphisms and growth); G. Sarap (dental anthroplogy).

RIGA
ACADEMY OF SCIENCES OF LATVIA,
INSTITUTE OF HISTORY
(226524, Riga, GSP, Rurgeneva, 19): R.Ya. Denisova (a leading specialist in racial history of the Baltic States; osteology and anthropometry); R.U. Gravere (dental anthropology).

VILNIUS
UNIVERSITY FACULTY OF MEDICINE
(232031 Vilnius, Ciurlionio 21): G. Cesnys (another leading expert in racial history of the East Baltic area). At the University Chair of Stomatology, dental anthropology is represented by I. Balciuniene.

MINSK
ACADEMY OF SCIENCES OF BYELORUSSIA,
INSTITUTE OF ART, ETHNOGRAPHY, AND FOLKLORE
(220600 Minsk, GSP, ul. Surganova, 1, korp. 2): L.I. Tegako (dental anthropology, dermatoglyphics), I.I. Salivon (anthropometry, growth), A.I. Mikulich (population genetics).

KIEV
ACADEMY OF SCIENCES OF UKRAINE,
INSTITUTE OF ARCHAEOLOGY
(252014, Kiev, Vydubetskaya, 40): S.I. Kruts (osteology), S.P. Segeda (dental anthropology).

TBILISI
ACADEMY OF SCIENCES OF GEORGIA,
INSTITUTE OF HISTORY AND ARCHAEOLOGY
(380000 Tbilisi, ul. Kamo, 51): M.G. Abdushelishvili (a leading figure in racial research of the Caucasus; anthropometry and osteology), V.F. Kashibadze (dental anthropology).

BAKU
ACADEMY OF SCIENCES OF AZERBAIJAN,
INSTITUTE OF HISTORY
(370143, Baku, Narimanova, 31): R. Kasimova (osteology).

ASHKABAD
ACADEMY OF SCIENCES OF TURKMENISTAN,
INSTITUTE OF HISTORY

(744000, Ashkhabad, Gogola, 15): O. Babakov (osteology).

ALMA-ATA
ACADEMY OF SCIENCES OF KAZAKHSTAN,
INSTITUTE OF HISTORY

(480000 Alma-Ata, Shevchenko, 28): O. Ismagulov (osteology, population genetics), K. Sikhimbayeva (dental anthropology, dermatoglyphics).

TASHKENT
UNIVERSITY FACULTY OF HISTORY

(700000 Tashkent): T.K. Khodjayov (osteology).

CHAPTER 3

NON-HUMAN PRIMATE RESEARCH IN THE USSR

M.L. BUTOVSKAYA

The Soviet school of primatology has a long history, and one of the characteristics of primatological investigation in the USSR is its interest in anthropological problems, especially with the study of human evolution. It is possible to differentiate four distinct periods in the development of primatological researches (Table 1).

THE FIRST PERIOD OF SOVIET PRIMATOLOGY

Soviet primatology began in the area of comparative psychology of apes. One of the first works on these topics was done by Ladygina-Kots (1923). Experimenting with chimpanzees, this investigator found that visual cues played the leading role in chimpanzees' orientation, while acoustics was of secondary importance. It was observed that chimpanzees were oriented primarily to the colour of objects, and were capable of differentiating approximately 30 variants of different colours. Chimpanzees were also able to differentiate objects by size and shape. The chimpanzees' preferences to bright variants of colours was interpreted by Ladygina-Kots as evidence of the existence of an esthetic awareness in their psychology. Similar to Köhler (1925), Ladygina-Kots had stressed the high level of chimpanzee abstract abilities. But contrary to him she noted that such abilities were the result of practical generalizations based on their own previous experience. Her main conclusion was that chimpanzees fulfilled tasks by using the principle of trial-and-error, but not by insight, as postulated in the works of Köhler.

In the early 1930s Ladygina-Kots had observed and compared the ontogenetic development of human and chimpanzee infants (Ladygina-Kots, 1935). It is interesting to note that her observations were made during the same time period as those of Kellog (Kellog and Kellog, 1933).

This early period of Soviet Primatology was characterized by an extremely important event which was the founding in 1927 of the first Primate Research Centre at Suchumi. This was the first institution in the world oriented to investigations in the field of medical primatology and primate breeding. The Laboratory of Comparative Physiology was set up at the Suchumi Primate Research Centre, and its basic projects were oriented toward observations of general characteristics of the animals' behaviour in groups, species-specific differences in group behaviour and experimental works on cognitive abilities of monkeys and apes. Special attention was paid to spontaneous investigatory and manipulatory abilities of monkeys (Voitonis, 1949). Voitonis had stressed that orientative-investigatory activity and a great degree of curiosity are basic peculiar features of primates' behaviour.

The works of Alexeeva (1948) and Tih (1948) started in the early 1940s at the Suchumi Primate Centre. Alexeeva's observations were oriented to the analysis of both physiological and behavioural changes in different stages of female sexual cycling. In harem groups of hamadryas baboons she noted changes in social status of females with swelling, and maximal spatial proximity to male leaders during this period. Alexeeva was the first to discuss the role of primate polycyclicity as an important prerequisite of human evolution (Alexeeva, 1948).

Tih's project was directed to a study of investigatory and manipulatory abilities of animals in groups, based on the comparative physiological approach. Basic mechanisms of group integration were examined. Tih concluded that sexual bonding is not the only factor that causes primates to live in groups. She stressed the importance of mother-infant relationships and individual orientations toward contacts and co-operation with conspecifics. The dominance-subordination relations are not limited to suppression of one individual by another, but also include protection of subordinates by those which are dominant, as well as subordinate individuals actively seeking such protection (Tih, 1948).

Also during this period studies on the physiology of primate behaviour were initiated at the Institute of Psychology (Leningrad). One of the first experimental works in this field analysed the question of the influence of the social environment on the individual's neural activity (disturbance of conditioned reflex reactions) (Dolin, 1936).

The level of chimpanzee intelligence was tested by Roginski (1948). He made experiments on teaching chimpanzees to communicate by means of chips, counters and Ameslan (American Sign Language). In his experimental tasks chimpanzees had to find a correct choice for food and play objects. Roginsky's works were done in parallel with experiments carried out by Cowles (1937).

The typical feature of primatological researchers of this first period was the use of Darwin's theory of natural selection as a cornerstone for understanding the evolution of psychology from primate to human level.

TABLE 1
THE MAIN STAGES OF DEVELOPMENT OF PRIMATOLOGY IN THE USSR

THEORETICAL BASIS	BASIC SPEHERES OF INTERESTS	BASIC BRANCHES OF PRIMATOLOGY
1917-1947		
• Darwin's theory of natural selection	Cognitive abilities of apes	Comparative psychology
	Modelling of human diseases	Medical primatology
1948-1964		
• Lamarkism	Sensory mechanisms, communication	Comparative psychology
• Pavlov's theory of conditional reflexes	constructive abilities, intelligence	
• Labour's theory of human construction	Morpho-biological characteristics of early stages of development of ape and human child	Ontogenetic studies
1965-1977		
	Intelligence, communication, social structure	Comparative psychology
	Memory and investigatory abilites	
• Darwin's theory of natural selection	Behavioural and physiological limits of adaptation	
• Mendel's theory of Inheritance	Brain Structure	Physiology of primate behaviour
• Pavlov's theory of conditioned reflexes	Physiology and behaviour of reproduction, communication in groups	
• Labour's theory of human evolution	Social roles and dominance relations	Social psychology
	Locomotion	Ontogenetic studies
1978-PRESENT		
• Theory of motivations, Synthetic theory of evolution	Behavioural parapmeters of stress, psychopharmacology and behaviour	Medical primatology
• Inheritance of locomotive acts	Neuronal activity, memory intelligence	Physiology of primate behaviour
• Theory of social psychology	Social relationships, communication, investigatory behaviour, individual behavioural differences.	Primate ethology and social psychology

All attempts of investigators were aimed at the search for similarities between the ape and human process of thinking, perception and on finding similar stages in the development of certain mental, communicative and sensory abilities of apes and man.

THE SECOND PERIOD IN SOVIET PRIMATOLOGY

The second period was the hardest one in the history of Soviet primatology, as well as for all Soviet natural science (Table 1). Mendelian genetics was strictly forbidden in the USSR and gradually all of Darwin's ideas about natural selection and Mendelian theory of inheritance were substituted by a primitive version of Lamarkism. Freedom of scientific thought became limited by ideological dogmatism. All findings which contradicted Marxist's paradigm about the leading role of labour in Man's origin were criticized. All investigations that supported the idea of close intellectual, psychological and behavioural similarity between apes and human were rejected, The search for similarities between human and primate cognitive and mental abilities, and ideas about gradual transformation of primate communities into human society were declared to be false. Pavlov's theory of conditioned reflexes became the single acceptable approach to interpreting all behavioural phenomena. Investigations in the field of physiology of primate behaviour became a major research area, and the main emphasis was placed on looking for qualitative differences in primate and human behaviour.

The work of Ladygina-Kots about constructive and tool-using behaviour of apes published during this period is one of the examples (Ladygina-Kots, 1959). Contrary to her first brilliant interpretation of complex cognitive phenomena in apes, she began to state that human intelligence is qualitatively unique and that the intelligence witnessed in tool-using activity of chimpanzees was of a basic elementary type. Chimpanzee mentality was described as having time-spatial relationships, but not causal-effectual ones characteristic in the transformation from a primate psyche to the human one. Ladygina-Kots insisted in this book that chimpanzees were not capable of tool-making.

Meanwhile, Pavlov himself was greatly impressed by the high mental level of apes. He was sure about an existence of a unique innate world in apes, and felt that chimpanzees possessed great cognitive potential expressed in their investigatory activities and curiosity. This founder of the theory of "conditioned reflexes" came, at the end of his life, to very important conclusions — that apes are able to grasp real connections between objects, and that constructive behaviour of chimpanzees cannot be interpreted purely by means of reflexes (Pavlov, 1949). However, Pavlov's followers seemed not to accept this idea. They continued to apply a "reflectory" explanation to all experiments on ape constructive behaviour, tool-using and memory (Shtodin, 1947; Schastny, 1972).

MOVING INTO THE THIRD PERIOD OF SOVIET PRIMATOLOGY

As changes came about in official ideology, the projects dealing with comparative psychology were curtailed. Only a few investigators continued to work in the field, but they (Tih, Alexeeva) were obliged to leave the Suchumi Primate Centre. The results of comparative studies of group behaviour of cercopithecoid monkeys (*Papio hamadryas, Macaca mulatta, Theropithecus gelada*) were published by Tih only in 1970. In her book this author talked about the existence of strong motivations for contacts with conspecifics, and she expressed the idea that competition and cooperation were different and independent forces acting in the group's everyday life. Tih also discussed the idea that hominid ancestors had possessed similar features of group behaviour. Her model of transformation from non-human primate communities to hominid societies was based on the idea of group selection. The motivating forces of transformation from ape-like ancestors to hominids was suggested to be collective provisioning and food distribution.

Tih showed that troop member cooperation outside of sexual bonds, and the increase in female relative freedom and free choice, were important in the evolution of hominids. Following the ideas of Kropotkin (1907) and Teillard de Chardin (1965), Tih concluded that brain development in hominid evolution was the product of modernization and complication of their social relationships, and she continued her investigations in the field of ontogenetic psychology, started earlier by Ladygina-Kots (1935). A major focus of Tih's work was the study of locomotive and sensitive abilities of monkeys, apes and the human newborn. Despite the number of inborn differences, she found similarity in processes of formation of the upper and lower extremities and various other features in apes and human (Tih, 1966). She insisted that both inborn and environmental factors were of equal importance for an individual's development.

During this third period of primate studies, Alexeeva expressed her ideas of the close connection between the development of social and sexual behaviour in primate evolution. She proposed a model of behavioural and physiological bases of anthropogenesis (Alexeeva, 1977). Firsov in Leningrad in 1972 began a project on ape acclimatization to its natural habitat. This was at practically the same time as a similar project begun at the Yerkes Primate Centre in Atlanta (Sellers, 1973) (Table 1). The main aim of both projects was to estimate adaptive abilities of apes both at the physiological and behavioural levels. In Firsov's project from 3 to 5 juvenile individuals were analyzed every summer season on the island in Pskovsky Region from 1972 to 1976. Provisioning was carried out only during the first two seasons; later it was stopped. Social relationships, abilities for long term memory, imagination and tool-using were tested in stimulus-obstacle testing (Firsov, 1977). All experimental works were carried out without isolation of individuals from other group members. Firsov presented data on ecology of feeding and a detailed list of plants used by chimpanzees on the island (a total of 116 plant species). Apes were shown to possess a high level of abstraction, fine memory abilities, the ability for application of old experience in new situations and the capability to operate by the use of symbols.

Another project on the acclimatization of hamadryas baboons was started at the Suchumi Primate Research Centre in 1972-1974. A troop of 76 wild monkeys was brought up from Ethiopia and released in the Gumista Reserve.

THE MODERN PERIOD OF SOVIET PRIMATOLOGY

Let me make a few general remarks before moving to a discussion of the characteristics of the modern period of primate research in the USSR:

1. It is important to stress the autonomous development of the Soviet School of Primatology. All previous studies in comparative psychology, ontogeny and the physiology of behaviour were developed in practically absolute isolation from contacts with Western colleagues.
2. Despite the rapid development of ethological and socio-ecological approaches in the West, by the end of 1970s these kinds of approaches in primate behaviour were still absent in the USSR.
3. Until the end of the 1970s all investigations of primate behaviour were of a naturalistic type, made in a descriptive manner. Any kind of quantitative methods was absent.
4. In the 1970s the only course of lectures on primate systematics (including behaviour) was terminated at Moscow State University. Thus for a certain period, none of the Universities in our country have prepared specialists familiar with primatological problems.

As a matter of fact, the last, modern period of primatological researches started approximately in 1978 with the restoration of a lecture course on "Primate systematics and behaviour." Since that time it has been regularly presented for the students of the Department of Anthropology (Moscow State University) by Dr. M.A. Deriagina in co-operation with Dr. M.L. Butovskaya (Table 1).

In 1978-1979 a group of specialists from the Department of Anthropology (Moscow) and the Suchumi Primate Research Centre combined their efforts on the project of development and application of an ethological structural-dynamic approach and modern ethological methods to studies of primates (Deriagina et al., 1984).

Contrary to classical ethology of the 1940s and 1950s which were oriented primarily to studies of instinctive behaviour and its mechanisms, the term "ethological" was used in a much broader sense to include investigations of complex forms of behaviour that are connected with long periods of learning. The structural-dynamic approach developed by these authors is based on the following principles:

1. Hierarchical organization of behaviour
2. Dynamics of behaviour
3. Quantitative registration
4. Systemic approach to the basis of the relationship of different forms of behaviour. (Butovskaya and Deriagina, 1989)

The principle of hierarchical organization of behaviour was proposed by Tinbergen (1942) and later developed by Panov (1978). According to these authors we have to distinguish different components of behaviour such as:

1. elementary locomotive acts;
2. postures and movements;
3. the level of sequential chains of postures and movements;
4. complexes of behavioural chains;
5. functional spheres.

From our point of view, the basic mechanisms of behaviour can be understood through studying the functioning of the highest levels of integration and the key level for understanding this is the level of sequential actions. The movements and their connection in the chain of actions determine such basic characteristics of behaviour as its degree of ritualization and variability, randomness and purposefulness.

Observation on the chain of actions provide information about active interactions of the focal animal with its environment, discrete objects and other individuals. The attribution of discrete postures and movements to certain functional spheres is made on the basis of the following principles:

1. estimation of the place of the discrete element in the chain of actions of the focal animal;
2. examination of the reaction of other group members;
3. situational analysis of each element;
4. estimation of the final aim of the chain of actions in which the discrete element is applied.

The basic principle of classical ethology, that is, stimulus-reaction, was used in this transformed way, because in most cases, elements in a system of stimulus-reaction are not simply connected. Concrete signals in many cases arise in a nondetermined way. This approach was tremendously useful in the work with primates, both because of the great degree of deritualization and the observation of polyfunctionality of elements of their behaviour.

The principles of classification and coding of elements of behaviour and basic methods of observations developed by an association of observers from Moscow State University and Suchumi Primate Research Centre (Deriagina et al., 1984) are used in the works of all Soviet scholars. Thus, the main results received by Soviet primatologists are comparable. Most observations are carried out by means of protocols and specially developed frequential, durational and correlation matrices (Butovskaya and Deriagina, 1989).

One of the first projects using this ethological approach was the comparative analysis of manipulatory and tool-using abilities in representatives of different primate taxa. The main purposes of this project were as follows:

1. to present an objective criterion of differences between tool-using activity of apes compared to other animals;
2. to find differences in the development of manipulatory and tool-using abilities in modern primate taxa;
3. to reconstruct the possible stages in the development of tool-using for the interpretation of human evolution (Deriagina, 1986).

The special set of tests with application of different objects was developed and characteristics of manipulatory abilities of 11 species were presented on the basis of 22 original indices, such as: the general number of variants of fixation, the number of variations of forms of manipulation (deformation, destruction, actions mediated with body), and indices. These characterize the complexity of the chains of manipulation. The basic conclusions of Deriagina (1989) were in accordance with ideas proposed by Parker and Gibson (1973), who used ecological and psychological criteria.

Three types of tool-using actions in animals were differentiated. The first one was the prototool-using behaviour. It was characterized by similarity of elements of manipulatory activity, primitive forms of fixation and short non-differentiated chains of manipulation with a great degree of stereotypic actions. The probability of the origin of new actions of such a type was thought to be very low. Examples of this type of manipulatory activity are found in capuchin monkeys and crab-eating macaques.

The second type, individual tool-using, is typical of a qualitatively higher level of development of manipulatory activity. It is characterized by differentiation of the ways of fixation, the growth of a number of forms of manipulation and complication of chains of manipulation. The tool-using actions of the second type are found in apes.

The third type, group tool-using, is characterized by complex chains of manipulatory activity, imitation and directional teaching. It is found only in chimpanzees and humans.

At about the same time as this project began, another one dealing with the comparative analysis of the structure and dynamics of aggressive behaviour was started. General and species-specific peculiarities of aggression and friendly behaviour in groups of 16 species were examined (Butovskaya, 1987). To estimate the possible ecological influence on the expression of aggression, aggressive patterns of behaviour were compared in groups of hamadryas baboons located in different captive conditions and in the Gumista Reserve (Butovskaya, 1984).

In 1982 studies on the social structure and ingroup social relationships began under the project "Biological prerequisites of anthropogenesis" granted by the Institute of Ethnography, Academy of Sciences of the USSR (Moscow). We approached the study of the social structure from the point of detailed analysis of proximal mechanisms of pair bonding, and from the point of ethological motivational mechanisms, that make individuals live in groups (Popov, 1986). Peculiarities of pair bond formation and the strength of such bonds were thought to be typical characteristics of species social systems. Without denying the positive role of the traditional socio-ecological approach for understanding the functioning of different social structures, we considered that the pair bonding phenomenon was not a consequence that came out of the necessity of life in groups. It was supposed that if such bonds were psychologically and emotionally positive and mechanisms of their fixation had already existed, then such bonds had to be expected even in situations of absolute absence of ecological benefits for individuals (Obsianikov, 1986). Investigation of social systems from the point of dyadic bonding gave us a chance to refer to the unique personality and not only as a representative of a certain age, sex or social status category. In this aspect our approach is close to Crook's "Socioecologically referred social psychology of personal behaviour and development" (Crook, 1989, p. 19). Interindividual social relationships were approached as unique phenomena, the result of complex influences of group history, social traditions, rank of birth, mother's positive and antagonistic relationships, individual's biography and emotional "portraits," and the concrete social and ecological situation at the time of bond formation (Butovskaya, 1989).

Interspecific differences in forms and intensity of aggression, the amount of reconciliatory and appeasement actions, as well as support in agonistic encounters and friendly activity and social permissiveness were examined on the basis of this approach (Butovskaya and Ladygina, 1989 a, b; 1990; Butovskaya, 1987; 1991).

For example, the role of kin-effect in the distribution of social bonds was analysed in groups of species from the genus *Macaca*. It was demonstrated that the sympathies and antipathies in the development of social structure of two groups of *Macaca arctoides* were important (Butovskaya and Ladygina, 1989, 1990). Comparative analyses of group social relationships in stumptail, rhesus, longtail and pigtail macaques showed evident differences in basic individual orientations. Social structure of stumptail macaques was found to be the most permissive, compared to three other species; rhesus and longtailed macaques were characterized by strictness and rigidity of dominant relations. In this aspect our conclusions are similar to findings of Thierry (1985) and to the idea of different "dominance styles" proposed by De Waal and Luttrell (1989).

The long-lasting project on acclimatization of hamadryas baboons in the Caucasus covered observations of demography and ecology. By the time of these observations, the wild population of hamadryas baboons in Gumista Reserve consisted of approximately 200 animals. Territorial behaviour and processes of harem formation were discussed in the works of Chalian with co-authors (Chalian and Meishwili, 1989; Chalian and Davidenko, 1989).

A project on the relationship beween dominant rank and reproduction in females of the two genera *Macaca* and *Papio* was realized under the support of the Suchumi Primate Research Centre and Institute of Ethnography, Academy of Science, USSR (Moscow) (Meishvili, et al., 1991; Chalian et al., 1991). Despite the fact that many authors had already tested socio-ecological predictions about reproductive success of females (Abbot 1987; Berman, 1985; Dunbar and Dunbar, 1977; Gouzoules, et al., 1982; Harcourt, 1987), the problem remained still far from being solved. We think that conclusions gathered from our studies should add information for a general understanding of the phenomenon.

It was found that low-ranking females were less successful (based on the number of surviving offspring) compared to high-ranking females (Meishvili et al., 1991). But such similar results in different species were attributed to radically different reasons. In the groups of longtailed macaques the highest number of births in low-ranking females was compensated for by a higher level of abortion and low rate of infant survival. In rhesus groups, where all infants have a generally high level of survival, a tendency for decreasing numbers of births and for a higher level of abortion in low-ranking females was found. Groups of stumptail macaques generally possessed a lower rate of births and infant survival compared to other species. In this species the most successful category of females were those in middle rank of the social hierarchy.

The aim of another project was to search for the reasons for infanticide in a group of hamadryas baboons, and to work out recommendations for reducing the probability of infanticide in captivity (Chalian et al., 1987). One hundred and sixty-five cases of infanticide occurred over a period of 20 years in captivity, and 14 cases — in a wild population from the Gumista Reserve — were discussed. It was concluded that infants of young primaparous females had the highest risk of being killed both in captivity and in the wild. Infanticide appeared to be most frequent in captivity in situations of the introducing of a new male leader. Similar results were reported by Angst and Thommen (1977). Chalian and co-worker agreed with Rijksen's (1981) hypothesis, that infant killing can be explained from the point of view of ingroup attention structure. New male leaders would fight for the attention of females who had inf ants younger than a month old. It was found that all infanticidal males were younger than 10 years of age and most cases of infanticide can be treated as a result of the male's frustration and lack of social experience. The victims of male aggression were not only infants of a former leader, but also his own offspring. Thus, contrary to sociobiologists (Hrdy, 1977; Hausfater, 1984) it was concluded that infanticide in hamadryas baboons cannot be explained as a variant of male reproductive strategy. The present state of the problem of infanticide in primate and human populations was recently reviewed by Butovskaya (1990).

The connections between social structure and manipulatory activity were analysed as part of a general project, "On the evolution of social structure and the origin of human society," granted by the Institute of Ethnology and Anthropology (Institute of Ethnography), Academy of Science USSR. For this purpose groups of stumptail and rhesus macaques were investigated in similar outdoor corrals. Social and manipulative "portraits" of all group members were collected (Butovskaya and Zagorodny, 1991). Manipulatory activities were observed both as spontaneous actions and in experimental sessions by the scheme proposed by Deriagina (1986). We have concluded that young and juvenile individuals possessed higher manipulative and investigatory potentials, while lactating females and individuals older than 15 years were the least active. Similar results were published for Japanese macaques (Huffman, 1986). The level of spontaneous manipulatory activity was independent of the individuals' social status. In the first stage of experiments, the high indices of manipulation, typical for high-ranking animals, was explained as consequences of social competition for desirable objects. Rank-dependent frequency of manipulation obtained for all-male groups of stumptails was interpreted to be a result of the usurption of objects by dominant individuals, as mentioned earlier by Kummer (1973). It was concluded that: 1) the development of manipulatory activity was closely connected with species-specific investigatory abilities; 2) in-group frequency of manipulation, broadness of such practice among group members, and possibility of accumulation and transition of invented manipulatory technique were found to be closely connected with group social structure. The climate in multi-male/multi-female social groups with well-developed matrilineal systems and mild social relationships was the most fruitful for the expression of tool-using; 3) psychological orientation of individuals towards cohesion with conspecifics was felt to be another positive factor that facilitates individual manipulatory behaviour and creates appropriate conditions of social status of the inventor (Butovskaya, 1989; Butovskaya, Zagorodny, 1991).

The project on investigations of the origin of language was granted by the Department of Anthropology of Moscow State University. It was concentrated on ethological and bio-acoustical observations of 11 primate species. A comparative analysis of gestural and mimic communication was performed and the general scheme of reorganization in communicative systems in primates' phylogeny was proposed (Deriagina et al., 1989). Firsov and Moiseeva (1989) have shown by their work that apes were able to operate by preverbal notions and thus, they already possessed a language system based on inborn communicative signals and on a gestural-sign system. In later works it was thought that language is a product of social association, but not a consequence of the origin of labour (Vasiliev and Deriagina, 1991). The development of language was viewed as a long-lasting gradual process but not due to a quantitative jump. The evolution of social complexity and erect bipedality were considered to be two main factors that stimulated the process of differentiation by means of communication.

SUMMARY

In this review I have concentrated on the basic steps of the development of Soviet primatology and of course, only a small portion of ideas and results have been mentioned. Soviet primatologists are ready for contacts with Western colleagues and are equally interested in mutual research, theoretical discussions on basic problems of primate behaviour and methodological approaches. I hope that this article will bring about greater interest in the works of Soviet primatology and will stimulate further scientific contacts.

*   ACKNOWLEDGMENTS

I am greatly obliged to Professor B. Sigmon and the administration of the Erindale College, University of Toronto for providing me with financial support for this review article and its presentation in the form of lectures both at the Symposium and at the Annual Session of Canadian Association of Physical Anthropologists. I am also grateful to Professor B. Sigmon for her assistance and recommendations on the manuscript.

REFERENCES

Abbott, D.H. 1987. Behaviourally mediated suppression of reproduction in female primates. *J. Zool. Found.* N. 13. P. 455-470.

Alexeeva, L.V. 1948. Materials of studies of primate sexual cycles in connection with anthropogenesis. PhD Thesis. Moscow. (in Russian).

————. 1977. *Polycyclicity of Reproduction in Primates and Anthropogenesis.* Moscow, Nauka. (in Russian).

Angst, W., D. Thommen. 1977. New data and a discussion of infant killing in old world monkey and apes. *Folia primatologica.* V. 27 P. 198-229.

Berman, C.M., R.C. Rawlinges. 1985. Maternal dominance, sex ratio and fecundity in the social groups on Cayo Santiago. *Int. J. Primatol.* V. 8 N.4. P. 421-423.

Butovskaya, M.L. 1987. The evolution of group behaviour in primates as premises of anthroposociogenesis. *Soviet Ethnography.* N 1. P. 52-69. (in Russian).

————. 1989. The reconstruction of group behaviour and social behaviour of early hominids by primatologic data. In: *Biological Prerequisites of Anthroposociogenesis* (V.R. Alexeev, M.L. Butovskaya (eds.)). V. 2. Moscow, Institute of Ethnography. P. 55-111. (in Russian).

————. 1990. The role of aggressive behaviour in the evolution of primate and early hominid societies. *Anthropologie.* V. XXYII/2-3. P. 249-253.

————. 1990. Infanticide in primates and the problems of reconstructions of hominids social hehaviour. *Biologicheskije nauki.* N. 12. P. 5-25 (in Russian).

————. 1991. The social relations in the group of stumptail macaques (Macaca arctoides) in connection with the change of the leader. *Primate report.* V. 29. P. 29-34.

————., M.A. Deriagina. 1989. The modern methods of investigations in primatology and the problems of anthropogenesis. In: *Biological Prerequisites of Anthroposociogenesis* (B. Alexeev, M.L. Butovskaya (eds.). V. 1. Moscow, Institute of Ethnography. P. 3-22. (in Russian).

————., Ladygina O.N. Support and co-operation in agonistic encounters of stumptail macaques (Macaca arctoides). *Anthropologie.* 1989. V. XXVII/1. P. 73-81.

————., O.N. Ladygina. 1990. The influence of kinship and individual dispositions in groups of stumptail macaques on social and sexual behaviour. *Bull. MOIP.* V. 95. N 6. P. 3-15 (in Russian).

————., O.N. Ladygina. 1990. Congenial relations and agonistic behaviour in groups of stumptail macaques. *Bull. MOIP.* V. 95. N 3. P. 3-19. (in Russian).

————., W.A. Zagorodny. 1991. The role of social factors in development of tool-using activity on the early stages of hominids evolution. In: *Primates Behaviour in the Problems of Anthropogenesis.* Moscow. Nauka. (E.Z. Godina (ed.)). P. 4-14. (in Russian).

Chalian, V.G., S.V. Davidenko. 1989. Acclimatization of hamadryas baboons in Black Sea Coast in Caucasis: territorial behaviour. *Biological Sciences.* N 4. P. 71-77. (in Russian).

————., N.V. Meishwili, M.A. Vančatova. 1987. Infanticide and primate evolution. In: *Behaviour as One of the Main Factors of Evolution* (V. Leonovičová, V.J.A. Novak (eds.)) Praha. P. 321-330.

————., N.V. Meishvili. 1989. Behaviour and stages of sexual cyclicity in hamadryas baboons as a model of intersexual interaction in early hominids. In: *Biological Prerequisites of Anthroposociogenesis* (Alexeev V.P., Butovskaya M.L. (eds.)) V. 1. P. 81-97. (in Russian).

————., N.V. Meishvili. 1989. Demographic characteristics of primate troop as a model of analogic structures in early hominids. *Soviet Etnography.* N 4. P. 115-122 (in Russian).

————., N.V. Meishvili, R. Dathe. 1991. Dominance rank and reproduction in female hamadryas baboons. *Primate report.* V. 29. P. 35-40.

Chardin P. Teillard de. 1965. The Phenomenon of Man. Moscow. Progress Publ.

Crook, J.H. 1989. Introduction: socioecological paradigms, evolution and history: perspective for the 1990s. In: *Comparative Socioecology* (V. Standen, R. Foley (eds.)). Oxford. Blackwell. P. 1-36.

Deriagina, M.A. 1989. Tool-using in primates: criteria and determination. Hypothesis of its development in anthropogenesis. In: *Biological Prerequisites of Anthroposociogenesis* (V. Alexeev, M.L. Butovskaya (eds.)). V. 2. Moscow, Institute of Etnography. P. 25-54. (in Russian).

————., M.L. Butovskaya, A.G. Semenov. 1989. The evolutionary transformation in phylogenesis of primates and hominids. In: *Biological Prerequisites of Anthroposociogenesis.* Moscow. Institute of Ethnography Acad. Sci. USSR. V. 1. P. 98-129 (in Russian).

————., V.G. Chalian, N.V. Meishvili, A.L. Artamonov, A.V. Sozinov, M.L. Butovskaya. 1984. The application of ethological methods in studies of primate behaviour. *Voprosy Antropologii.* V. 73. P. 128-135 (in Russian).

De Waal, F.B.M., L.M. Luttrell. 1989. Toward comparative socioecology of the genus Macaca: different dominance styles in rhesus and stumptail monkeys. *Amer. J. Primatol.* V. 19. N 1. P. 83-109.

Dolin, A.O. 1936. Analysis of some biological moments, which change the high neural activity of animals. *Physiological J. of the USSR.* V. 21 N 4. P. 582-607. (in Russian).

Dunbar, R.J., E.P. Dunbar. 1977. Dominance and reproductive success among Gelada baboons. *Nature.* N. 266. P.351-352.

Firsov, L.A. 1977.*The Behaviour of Apes in Natural Habitat.* Leningrad. Nauka. (in Russian).

————., L.A. Moiseeva. 1989. Memory as a factor of anthroposociogenesis. In: *Biological Premises of Anthroposociogenesis* (Alexeev V.P., Butovskaya M.L. (eds.)). V 2. P. 3-24. (in Russian).

Gouzoules, H., S. Gouzoules, D. Fedigan. 1982. Behavioural dominance and reproductive siccess in female japanese monkeys Macaca fuscata. *Anim. Behav.* V. 30. N 4. P.1138-1150.

Harcourt, A.H. 1987. Dominance and fertility amoung female primates. *J. Zool. Found.* V. 213. N 3. P. 471-487.

Hausfater, G. 1984. Infanticide: comparative and evolutionary perspectives. *Current Anthropol.* V. 25. N. 4. P 500-502.

Hrdy, S.B. Infanticide as a primate reproductive strategy. *Amer. Sci.* V. 65. N 1. P. 40-43.

Huffman, Quiatt D. 1986. Stone handling by japanese macaques (*Macaca fuscata*): implications for use of stones. *Primates.* V 27. N 4. P. 413-423.

Kellog, W.H., L.A. Kellog. 1933.*The Ape and the Child: The Study of Environmental Influence Upon Early Behaviour.* N.Y., London. Kohler, W. *The Mentality of Apes.* N.Y., Harcourt, Brace. 1925.

Kropotkin, A.P. 1907.*Mutual Support as the Factor of Evolution.* Moscow. Znanie Publ. (in Russian).

Kumme, H. 1973. Dominance versus possession. An experiment on hamadryas baboons. Symp. IY Int. Congr. Primatol. V. 1. *Precultural Primate Behaviour.* P. 192-225.

Ladygina-Kots, N.N. 1923. The investigation of cognitive abilities of chimpanzee. In: *Materials of Zoopsychological Laboratory of Darwin's Museum.* Moscow, Petrograd. (in Russian).

————., N.N. 1935.*Chimpanzees Infant and Human Child.* Moscow. (in Russian).

————., N.N. 1959. *The Constructive and Tool-making Abilities of Apes (chimpanzees).* Moscow. Acad.of Sciences USSR. (in Russian).

Meishvili, N.W., M.L. Butovskaya, W.G. Chalia. 1991. Dominant rank and reproduction in females macaques. In: *Primates Behaviour and the Problems of Anthropogenesis.* Moscow. Nauka. (E.Z. Godina (ed.)). P. 26-38. (in Russian).

Meishvili, N.V., V.G. Chalian, M.A. Vancatova. 1991. Mother-infant relationships in Cynomolgus monkeys (*Macaca fascicularis*). *Primate report.*. V. 29. P. 41-46.

Obsianikov, N.G. 1986. Biographical method in investigations of mammal's populations. In: *Methods of Investigations in Ecology and Ethology.* Pushino. P. 157-171.

Panov, E.N. 1978. *Mechanisms of Bird's Communication.* M. Nauka. (in Russian).

Parker, S.T., K.R. Gibson. 1973. The importance of the theory for reconstruction the evolution of language and intelligence in hominids. In: Symp. IV Int. Congress Primatol. V. 1. *Precultural Primate behaviour.* P. 43-64.

Pavlov, I.P. 1949. *Pavlovskije sredy.* V III. P. 262-263. (in Russian).

Popov, S.V. 1986. Social interaction and social structure the possible connections of behavioural characteristics and populational structures. In: *Methods of Investigations in Ecology and Ethology.* Pushino. P. 121-140 (in Russian)

Rijksen, H.D. 1981. Infant killing: a possible consequences of a desputed leader role. *Behaviour.* V. 78. N 1-2. P. 138-161.

Roginski, G.Z. 1948. *Habits and Premises of Intellectual Actions in Apes (chimpanzees).* Leningrad. (in Russian).

Schastny, A.I. 1972. *The Complex Forms of Apes Behaviour: Physiological Investigations of "Spontaneous" Activity of Chimpanzees.* Leningrad, Nauka. (in Russian).

Sellers, T. 1973. The esland apes. *Yerkes Newsletter.* V. 10. N. 1. P. 21.

Shtodin, M.P. 1947. About some forms of apes behaviour in experiments. In: *The Works of the Institute of Evolutionary Physiology and Pathology of High Brain Activity* named by I.P. Pavlov. V. 1. Academic Publ. P. 191-200. (in Russian).

Tih, N.A. 1948. Troop relationships in monkeys and communication in the light of problem of anthropogenesis. Ph.D. Thesis. Moscow. (in Russian)

_____. 1966. *The Early Ontogenesis of Primate Behaviour.* Leningrad. Leningrad State Univ, Publ. (in Russian).

Tinbergen, N. 1942. An objectivistic study of the innate behaviour of animal. *Bibl. Biothr.* V. 1. N. 1. P. 39-98.

Thierry, B. 1985. Patterns of agonistic interactions in three species of macaques (*Macaca mulatta, M. fascicularis, M. tonkeana*). *Aggressive Behaviour.* V. 11. P. 223-233.

Vasiliev, S.W., M.A. Deriagina. 1991. The forms of communication in primates and the stages of the origin of language. In: *Primate behaviour and the problems of anthropogenesis.* Moscow. Nauka. (E.Z. Godina (ed.)). P. 15-25. (in Russian).

Voitonis, N.U. 1949. *The Prehistory of Intelligence on the Problems of Anthropogenesis.* Moscow-Leningrad Academic Publ. (in Russian).

CHAPTER 4

GENETIC ADAPTATION STUDIES IN THE SOVIET UNION: MALAISE AND CURE

R.I. SUKERNIK AND M.H. CRAWFORD

INTRODUCTION

The concept of adaptation has been used in the literature to describe various facets of genetic-environmental interactions in human populations (Dubos, 1965). For example, cultural anthropologists have utilized this term to represent a general (usually unidentified) response to environmental or societal stresses. Geneticists define adaptation as a long-term, populational genetic response to environmental stress through natural selection. In order to eliminate further ambiguity, Lasker (1969) summarized the various uses of this term and suggested that the term "adaptability" be utilized generally, while adaptation be restricted to a populational response through selection. He also identified "acclimatization" as a reversible, short-term physiological response by the organism. He termed "plasticity" any irreversible morphological response of an organism through the growth process to some environmental stress, such as hypoxia.

In this paper, we limit ourselves to the concept of genetic adaptation. In particular, we consider the relative importance of adaptive and non-adaptive changes that have had a pervasive impact on the genetic structure and evolution of relatively isolated human populations of Siberia. Most population studies in the USSR, involving human genetic adaptation to harsh environments, have been conducted in the Siberian Arctic. Figure 1 locates geographically the major indigenous populations of Siberia that are mentioned in this chapter. Genetic variation observed in these indigenous groups has been summarized by Szathmary (1981) and Spitsyn (1985).

FIGURE 1
THIS MAP OF SIBERIA, WEST OF THE URAL MOUNTAINS, GEOGRAPHICALLY LOCATES THE MAJOR ETHNIC GROUPINGS RESIDING ON THAT EXPANSIVE LAND MASS

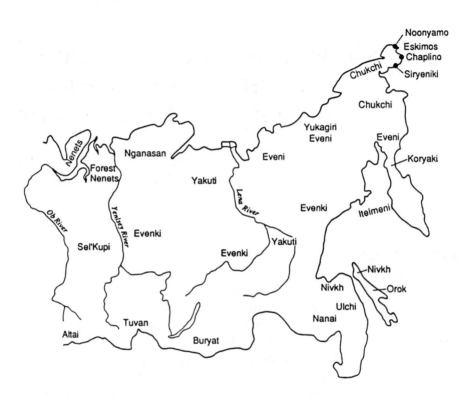

Historical Background

Since genetic adaptation operates through natural selection, the concept of fitness (W) as measured by differential fertility must be taken into consideration. Fitness includes both the struggle for existence between individuals comprising populations of the same species and the organismic interaction with the environment. Adaptation to the environment may be achieved through genetic specialization, phenotypic plasticity, or by the interaction of both. Genetic diversity and phenotypic modification are complementary, rather than alternative methods of adaptation (Dobzhansky, 1969; Cavalli-Sforza and Bodmer, 1971). Crawford (1987) has suggested the existence of a relationship between extreme genetic diversity and the successful genetic adaptation of colonizing Black Carib populations. The triracial hybrid populations provide an evolutionary success story as a result of their high genetic and phenotypic variation which permitted their expansion into a series of environments along the coast of Central America. Similarly, the ethnohistory of Siberia reveals massive genetic admixture resulting from gene flow between the indigenous groups and hybridization with Russian settlers. Thus, Siberia is an excellent location for applying dynamic models of genetic admixture and heterozygosity in the understanding of genetic adaptation.

By the mid-1920s, Russian geneticists explored the concept of genetic adaptation. For example, S.S. Chetveryakov (1927) suggested the concept of genetic homeostasis as an adaptive mechanism through which the population maintains its genetic structure despite the influence of environment. N.P. Dubinin (1931), simultaneously with Sewall Wright, presented genetic drift as an nonadaptive force, combining in it all stochastic factors. Drift was not considered as a rival or alternative to natural selection. Dubinin theorized that a combination of both drift and selection operated on mutations and their interaction was absolutely necessary for the action of "organic evolution." More recent developments in population studies and molecular genetics confirm the original views of Dubinin and Chetveryakov that selection and drift interact in a long-term process of trial and error that results in genetic adaptation.

The Malaise

Prior to Stalin's 1935 purge, genetics in the Soviet Union was developing as a permanent and successful branch of biology and medicine. Two internationally known genetic scientific communities were developing in Moscow and Leningrad. An Institute of Medical Genetics was established in Moscow in 1929. N.I. Vavilov, founder of the school of modern biology and genetics of the Soviet Union, contributed much to the concept of genetic adaptation. He considered the abundance of genetic diversity within the place of origin of any species as a major prerequisite for its successful spread and adaptation in the periphery of the area with a harsh envirnment.

As early as 1935, political attacks were launched against the so-called "formal" or "Mendelian" genetics school. These attacks were led by academician Trofim Lysenko, a Lamarckian agricultural biologist who was promoted by Stalin to head the rapidly bureaucratized biological sciences. Lysenko imposed Lamarckian concepts on the predominance of nurture over nature in the process of creating new forms with desirable characteristics. He promised to feed the country by creating new sorts of plants and new breeds of domesticated animals, more productive than the old ones and within the span of one or two generations. These Lamarckian concepts were so seductive to the communist government because they better fit the prevalent ideology. The most ambitious goal of the Soviet leaders was to develop a method to create a new type of person who would be appropriate for the building of communism. Since this officialdom considered equality not only as a sociopolitical but also a biological concept, Lysenkoism became an instrument of ideology. Even in the 1970s in some Soviet medical schools and universities, students were instructed that most quantitative traits, including human behaviour, could be modified within the individual's life and this change depended mostly on social conditions, almost irrespective of genetic legacy. In accordance with Lysenko's doctrine, any individual could be moulded into some sort of a "worker bee" that could be sacrificed for the sake of the happiness or well-being of future generations, who in turn would live in a Utopian Communist Society.

The official acceptance of Lysenkoism by the Soviet government had immediate and also long-term repercussions on genetics. Both the Leningrad and Moscow schools of genetics were totally destroyed and the internationally known scientists either perished or were sent into exile. In the late 1950s, a slow rebirth in research and teaching of basic genetics took place in a few institutions. For example, the "old" geneticists headed first by N.P. Dubinin and then by D.K. Belyaev gathered in Novosibirsk under the umbrella of the Siberian Branch of the Soviet Academy of Sciences. This group did not include human genetics, which was still under suspicion and continued to be so even after the formal denounciation of Lysenkoism in the early 1960s. The reason for the exclusion of human genetics involves the concept that everything involving human nature should be under the supervision of the highest political authorities and by the nomenclatura that it created. Until recently, the major concepts of human genetics remained at odds with official Soviet ideology and its study was discouraged. These negative historical events led to a retardation of Soviet human genetic research, its teaching and the practice of medical genetics. In fact, the whole area of human biology is underdeveloped in the Soviet curriculum. The opportunities for the development of human genetics were relatively bleak until a few years ago when some life was injected into current research within the frame of the Human Genome Project.

The history of modern biology and genetics in the Soviet Union vividly illustrates how ideology can interfere with scientific methodology. Lysenkoism had devastating consequences on the development of genetics in the USSR and its pernicious effects can still be seen.

STUDIES OF GENETIC ADAPTATION

The earliest application of genetic theory to human populations of the USSR can be attributed to Yuri G. Rychkov from Moscow State University. He applied a series of complex models to a collection of gene frequency data representing several blood group loci from indigenous populations of Siberia (Rychkov et al., 1969; Rychkov and Sheremetyeva, 1972; 1977). His ultimate research goal was the measurement of the action of natural selection, in order to isolate the effects of adaptive factors in various regions of Siberia. Rychkov attempted to measure the effects of selection by comparing the equivalent of Wright's F_{st} values between loci and by examining deviations from Hardy-Weinberg expectation. Unfortunately, many of the samples employed were inadequate, both in relation to size and their ethnic composition. His method of "painting the genetic landscape" of Siberian populations with broad brush strokes loses much detail and glosses over some of the problems of the original data sets. The problems included: (1) insufficient sample sizes of dubious origins; (2) technical errors, apparently present in the data. Thus, the incidence of alleles and haplotypes at the Rhesus, Duffy and Diego loci, as reported by Rychkov and Sheremetyeva for the Chukchi and Eskimos, appears to be at odds with the literature. For example, Ferrell et al. (1981) demonstrated that the probability of obtaining the FY*A allelic frequency at the Duffy locus given by Rychkov and Sheremetyeva (1972) was in fact 1×10^{-53}, (3) Rychkov's failure to consider alternative explanations of the results obtained. A more parsimonious explanation of the action of selection for the observed deviations appears to be sampling error or the action of stochastic processes, or both. Nevertheless, Rychkov's research was innovative for its time and place and set in motion many other projects.

Another focus of investigations devoted to the genetic adaptation of both human and other species was developed by Yuri Altukhov, Moscow Institute of General Genetics of the Soviet Academy of Sciences. Altukhov formulated the concept of "adaptive norm" which he defines as:

> "The principle prerequisite for persistence, stability and reproduction of natural populations is optimal genetic variation. Both its increase and decrease, usually taking place under the influence of diverse environments, is equally unfavorable for the normal functioning of the populations." (Altukhov, 1989, p.5)

Following in these footsteps, Yuri E. Dubrova and his colleagues have attempted to correlate genetic heterozygosity at 12 genetic loci with fertility in women who have completed their reproductive careers (Dubrova et al., 1990). In this way they had hoped to demonstrate that selection favours an optimal degree of genetic variation. The genetic and fertility data sets were collected from Forest Nentsi (Nenets) and the Nganasan by R.I. Sukernik's research group from Novosibirsk. The existence of a U-shaped curvilinear relationship between the index of achieved reproduction (a measure of reproductive success) and mean per locus genetic heterozygosity (a measure of genetic variation) in both tribes is suggested by Figure 2. Dubrova et al. conclude that this relationship indicates optimum fitness of average genetic heterozygosity and is evidence for the action of normalizing selection against extreme genetic variability.

FIGURE 2
THESE TWO FIGURES ARE PLOTS OF THE RELATIONSHIP BETWEEN THE LEVEL OF INDIVIDUAL HETEROZYGOSITY (ABCISSA) AND THE AVERAGE NUMBER OF PREGNANCIES OF THE FOREST NENTSI (A) AND NGANASAN (B).

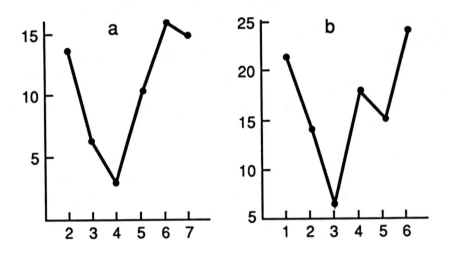

Unfortunately, this particular study was based upon small and somewhat heterogeneous samples that had to be subdivided into at least six miniscule and unequal categories for analysis. In addition, this estimate assumes an equal distribution of individuals into different categories of variation. Actually, the greatest concentration of this sample is in the average heterozygosity category with a few outliers representing the extreme variation. Apparently, this U-shaped curve merely represents a normal distribution of heterozygosity classes rather than demonstrating the action of stabilizing selection.

The concept of the "adaptive norm," resurrected in the USSR by Altukhov and Dubrova, has historical precedents. Lerner (1954) first developed the concept of genetic homeostasis which is based upon the relationship between selection and phenotypic variability (for a review of this topic see Dobzhansky, 1969; Chakraborty et al., 1986). Ostensibly contradictory results were obtained from a variety of traits such as growth rate, morphological variance, developmental stability, environmental variation and behaviour. A number of publications report the operation of normalizing selection in human populations. Schmitt et al. (1988) have argued that heterozygous individuals express their genotypes more optimally and as a result show less morphological variation than do the homozygotes. However, the research on the Aymara by Chakraborty et al. (1986), Black Caribs by Crawford and Comuzzie (1989) and Mennonites by Comuzzie and Crawford (1990) failed to support this purported relationship between morphological and genetic variation. Thus, conclusions by Dubrova et al. (1990) regarding selection and genetics have not been substantiated by studies on the Aymara Indians, Black Caribs and Mennonites.

Tatyana I. Alexseyeva (Institute of Anthropology, Moscow State University) has been investigating the adaptive processes in human populations, particularly those relating to ecology and organismic physiological response. In her "anthropoecological" surveys of Siberia and other rural areas of the Soviet Union, Alexseyeva deals with a series of morphological traits (mostly linear) that tend to exhibit a higher genetic component than the circumferential measures of the human body. These measures include stature, weight, chest width, upper arm length, leg length, head length, head width, systolic and diastolic blood pressure, blood protein, gammaglobulin, cholesterol and an assortment of other variables. She attempted to establish correlations between these parameters and the composition of the soil, climatic conditions and other ecological variables. Based upon these associations, Alexeyeva suggested that Siberian people belong to specific adaptive types irrespective of their ethnicity and racial origin. She defines an "adaptive type" as a standard biological response to the environment that ensures the state of relative equilibrium of the population with the environment. This adaptive type is expressed in specific morphological variation in populations. Adaptive types apparently derive from converging processes that take place in populations of different ethnic origin, through residence in similar habitats (Alexseyeva, 1989; 23). The use of the term adaptive type is suggestive of the action of natural selection on each population in favor of specific traits that permit them to better withstand the cold stress. However, this approach neglects the possible environmental stresses that may affect the growth and development of the organism. Judging from the high altitude studies in the

Andes and the Himalyas, the more parsimonious explanation for observed common phenotypes at such environments is plasticity rather than genetic adaptation. Severe hypoxia, cold, and heat may effect the growth processes which in turn modify the morphology of migrants. These changes are brought about through genetic/environmental interaction operating on the growing organism rather than resulting from genetic adaptation.

For almost two decades, V.A. Spitsyn (Institute of Medical Genetics, USSR Academy of Medical Sciences, Moscow) has been studying genetic variation of blood protein markers in various Siberian ethnic groups. He has detected some correlations between the distribution of gene frequencies and geography. He has claimed that the highest frequency of red cell acid phosphatase 1 A allele (ACP1*A) in the Eskimos was due to stabilizing selection (Spitsyn, 1985). Geographic regularities in the distribution of this and other genetic markers in relation to longtitude/latitude, described by Spitsyn, are in agreement with other studies reported in the literature. However, non-adaptive processes, such as massive historical movements play an important role in the observed patterns of variation in contemporary Siberian populations. Unfortunately, the action of migration and gene flow in these patterns has been ignored.

Recently, V.A. Spitsyn shifted his interest somewhat and at present is searching for associations between the same discrete traits at the protein level and a variety of environmental, ecological and pathological parameters (Spitsyn and Novoradovsky, 1989). Unfortunately, no synthetic models have been developed that would approximate the interaction between ecology and genetics. Given sufficient attempts at correlation between a large number of genetic and ecological variables, some statistically significant relationships will be observed by chance alone. Correlations between 20 or more variables will produce at least one significant r value at the 0.05 level of probability.

The research programs developed by V.P. Pusyrev at the Tomsk Institute of Medical Genetics, Siberian Branch of the Soviet Academy of Medical Sciences, and by L.L. Solovenchuk at the Magadan Institute of Biological Problems of the North, Soviet Academy of Sciences, echo the ideas and methodology of Spitsyn and Novoradovsky (1989). Solovenchuk states that the "genetic structure of newly formed populations of Chukotka (composed mainly of Russian-speaking immigrants) turns out to be similar to that which is unique to the native peoples of this region." The belief that specific genes at structural loci such as acid phosphatase are a prerequisite for a successful adaptation to the harsh environment of Northeastern Siberia is shared by a number of researchers (Spitsyn and Novoradovsky, 1989; Alexeyeva, 1989). They believe that migrants into the Arctic region exhibit similar gene frequencies as seen among the natives. The interesting question is how do the frequencies of the migrants reach the levels of aboriginal populations of Siberia? If selection is involved then there must be exceptionally high selection coefficients to account for the shift in gene frequencies within one or two generations. To date, fitnesses and selection coefficients of such high magnitudes have not been observed in Siberian populations. The ecogenetic ideas of Solovenchuk (1979; 1984), serve as illustrations of the consequences of Lysenkoism on the research of scholars during the last two generations. Many scientists raised in the grip of the officially accepted deterministic ideology of Lysenko continue to espouse theories that are tainted by Lamarckian and Communist philosophy.

INDIRECT ASSESSMENT OF SELECTION INTENSITIES IN SIBERIAN NATIVE POPULATIONS

Methodologically, it is difficult to directly assess the intensity of natural selection in human populations. The selection coefficients (S) and fitness (W) have to be of extremely high magnitude before population sampling can be partitioned from differential fertility. During the 1950s and 1960s, the discovery of hemoglobin S and its relationship to falciparum malaria, affected our view of genetic adaptation. Much like the "quest for the holy grail," similar balanced polymorphic models were sought at the other blood marker loci and with few exceptions these excerises proved futile. There are few successful direct measurements of selection operating on human populations. Examples include a number rare genetic diseases such as hemophilia (Haldane, 1946) and congenital achondroplastic dystrophic dwarfism (Morch, 1941), and more common phenotypes: birth weight extremes (Karn and Penrose, 1951), erythroblastosis foetalis and the Rhesus blood group system (Levine et al., 1941), and a number of genetic loci involved in providing resistance from malarial parasites (such as hemoglobins S, C and E (Allison, 1961), G6PD deficiency (Motulsky, 1960) and the Duffy FyFy phenotype (Miller et al., 1976).

Since selection operates through differential fertility and mortality, James Crow (1958) proposed a demographically based index for measuring the opportunity for selection in human populations. This index includes two components: 1) fertility I_f, which is made up of V_f / X^2, with V_f being the variance in the number of progeny per parent, X^2 is the mean number of progeny of women who have completed their reproductive careers. 2) mortality $I_m = p_d / p_s$; which is a ratio of the percentage dead and survivors. The total index $I_t = I_m + I_f / P_s$, is a rough measure of the opportunity for selection. Being a property of a population, Crow's Index provides no information as to what portion of the variance in mortality or fertility is genetic or can be assigned to a specific cause (Crow, 1989). However, it is a useful indication of which populations, subdivisions or temporal units may be experiencing demographic conditions that are conducive to rapid evolutionary change (Crawford and Goldstein, 1975).

Table 1 summarizes the components of Crow's Index, the average inbreeding coefficients and the mean per locus heterozygosity for four

Siberian indigenous populations. Three of these four Siberian groups were surveyed by the Novosibirsk research group at the Institute of Cytology and Genetics. The proportion of premature deaths was close to 50%, while the mean number of live births per woman did not exceed 7.3. Crow's total index was highest in the Nganasan, Eskimos and Forest Nentsi when compared to the geographically isolated Evens residing in Yakutia. The Evens, in contrast to the other three groups, have been expanding numerically during the last generation. This increase in population size can be explained by a comparatively higher standard of life resulting from their reindeer herding and breeding (Posukh et al., 1990). An alternative explanation for the evolutionary success of the Eveni through higher fertility and lower mortality is genetically determined fitness. If genetic heterosis plays a major role in the success of Arctic populations, then the Forest Nentsi should be numerically expanding, since they possess the highest level of heterozygosity.

TABLE 1
FERTILITY, MORTALITY, OPPORTUNITY FOR SELECTION IN COMPARISON WITH THE AVERAGE INBREEDING COEFFICIENT AND PER LOCUS HETEROZYGOSITY

INDICES	POPULATIONS			
	FOREST NENTSI (64)	NGANASANS (83)	EVENS (83)	ESKIMOS (91)
X Mean number of live births per woman who completed her reproduction	6.92	7.29	7.18	7.03
Vf Variance in offspring number	9.63	9.86	13.52	7.61
Ps Proportion of individuals that survive to an age of 15 years	0.56	0.46	0.76	0.49
Pd Proportion of premature deaths	0.44	0.54	0.24	0.51
If Index of potential selection due to fertility (If = Vf/X^2)	0.20	0.18	0.26	0.15
Im Index of potential selection due to mortality (Im = Pd/Ps)	0.79	1.17	0.32	1.04
It Total index of potential selection (It = $Im + If/Ps$)	1.15	0.3232	0.313	0.307
α Average inbreeding coefficient	0.015	<0.003	0.000	0.000
H Average Heterozygosity*	0.368	0.323	0.313	0.307

* Estimated for 12 loci: ABO, MNSs, P1, Rhesus, Duffy, Diego, Hp, Tf, 6-PGD, PGM1, AcP, Gm

Table 1 also provides the average inbreeding coefficient and the average level of heterozygosity for the four Siberian indigenous populations. The inbreeding coefficients were computed by Wright's pedigree pathway method (Wright, 1922) and indicate a moderate level of inbreeding in contrast to its absence in the Siberian indigenous groups. The Forest Nentsi have the highest inbreeding coefficient because of their frequent practice of consanguineous matings. The other three populations practise exogamy, and all four of these groups have considerably lower inbreeding than the levels observed in small genetic isolates like Tristan da Cunha and the Samaritans of Israel (Reid, 1973). Despite the differences in inbreeding levels, the mean per locus heterozygosities are relatively similar in these populations.

The Eskimos of Naukan and Chaplino have been displaced from their traditional sea mammal hunting areas. This forcible relocation, together with malnutrition, alcohol abuse, and disease epidemics has caused the rapid fragmentation of their genetic structure. Intermarriage and rapid gene flow with Chukchi and Russian settlers has given rise to a highly heterogenous gene pool and has all but obliterated the distinctive Eskimo population structure. Thus, Siberian gene pools have been rapidly changing as a result of admixture rather than through adaptation to the Arctic environment. Figure 3 compares the population pyramids of four Siberian Eskimo settlements. The dark portions of these pyramids represent non-Eskimo components and demonstrate the rapid replacement and hybridization experienced by these native groups.

FIGURE 3
THIS IS A COMPARISON BETWEEN THE POPULATION PYRAMIDS OF FOUR ASIAN (SIBERIAN) ESKIMO SETTLEMENTS. THE STIPLED PORTIONS OF THE COHORTS REPRESENT NON-ESKIMO RESIDENTS OF THE COMMUNITIES.

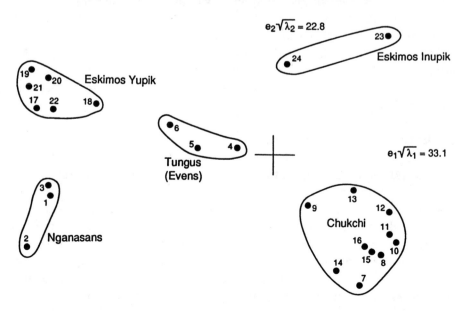

RARE GENETIC VARIANTS AND GENETIC STRUCTURE

Almost every Siberian genetic isolate, when surveyed genetically, exhibits unique or private genetic markers. For example, two unusual, deleted IGHG (gamma globulins as designated by the International System for Human Gene Nomenclature, ISGN, 1987)) haplotypes were found in near polymorphic frequencies in two adjacent Samoyed-speaking tribes, the Sel'kups and the Forest Nentsi. One of these markers, GM*-NB was localized in the Forest Nentsi of Pur River basin, while the other deletion, GM*-N'G occurs within the boundaries of eastern Sel'kups residing between the Pur and Yenisey rivers (Sukernik et al., 1991). Another rare genetic marker is 1F-1A2 allele at the Group Specific locus (vitamin D-binding protein) recently observed in the sera of Naukan Eskimos (Posukh et al., unpublished observations). These "private polymorphisms" as defined by J.V. Neel (1989) occur in tribal populations with intact genetic structure and are essential for the understanding of the relative roles of selection and drift in the evolution of tribal populations. Although Neel's fission-fusion model is more appropriate for Amerindian tribes of the tropical rain forest, major features of this model can readily be seen in Siberian indigenous populations. Siberian reindeer hunters such as the Nganasan, fishermen like the Forest Nentsi, and nomadic reindeer herders like the Chukchi and Evens, experience population subdivision and fusion at various times (Goltsova and Sukernik, 1979; Sukernik and Crawford, 1984). The sum total of the evolutionary processes operating on the populations and their subdivisions can be summarized by so-called "genetic maps." These topological plots are based upon principal components analyses of gene frequency variance-covariance matrices and provide a visual representation of population affinities and genetic structure (see Figure 4). This figure shows the clustering of population subdivisions of 24 Arctic groups reflecting their evolutionary history and possible genetic adaptation. The Samoyed-speaking Nganasan, situated on the Tamyr Peninsula appear to be the most unique of the Siberian groups, separated along the first eigenvector.

GENETICS, ECOLOGY AND HEALTH

If present, Mendelian disorders are extremely infrequent in Siberian Arctic and Subarctic populations. It is not surprising that there is an absence of autosomal recessives in small-sized aboriginal groups with a history of considerable duration. These observations are in agreement with basic population theory that recessive deleterious alleles should accumulate in large outbred populations. Some Siberian groups, such as the Forest Nentsi and the northern Altayans, practise consanguineous matings (see Table 1). The kinship structure of most of the other populations, such as the Nganasans, is based upon bilineal exogamy. During the history of these small inbred populations, selection prevents the accumulation of these deleterious alleles.

FIGURE 4
THIS IS A "GENETIC MAP" BASED UPON R-MATRIX ANALYSIS THAT SHOWS THE GENETIC AFFINITIES BETWEEN 24 POPULATION SUBDIVISIONS OF FIVE ETHNIC GROUPS OF SIBERIA

Asian Eskimos

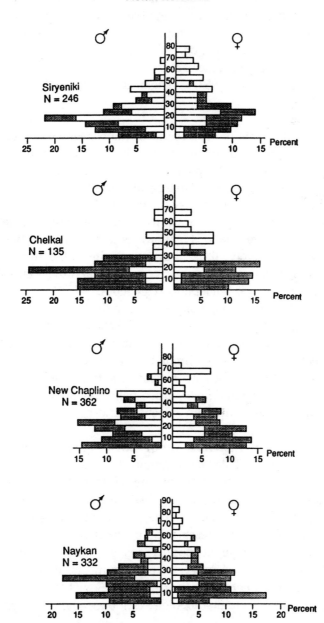

One interesting finding among the Forest Nentsi was the presence of congenital hip dislocation. This anomaly occurs at more than 30 per 1000 in Nentsi women, the second highest incidence in the world. The Lapps have the highest frequency and, like the Nentsi, they used to tightly swaddle infants and utilize portable cradles. It is believed that these cultural practices could exacerbate hip dislocation in predisposed infants. Females with narrow pelves and shallow acetabulum are of particularly high risk of exhibiting hip dysplasia. This condition provides an example of a complex interaction between a susceptible genotype, environment and cultural traditions to produce a high incidence of a genetic anomaly. Congenital hip dislocation also occurs in indigenous populations of the New World with a high incidence among S.W. and Canadian Amerindians (Kraus and Schwartzmann, 1957; Corrigan and Sega, 1950).

THE CURE

The first part of this paper describes the consequences of ideology dictating scientific methodology and theory in the Soviet Union. The results of Lysenko's ascendency to power were the persecution of highly innovative geneticists, the destruction of two major centres of genetic research and the training of several generations of geneticists and anthropologists with latent Lamarckian leanings. Although Soviet genetics and anthropology have made great strides during the last two decades, there has been relatively little creative activity in the area of human genetics. The last part of this paper considers possible "cures" for the "malaise" that has afflicted Soviet human genetics.

The comingling of ideology and science in the Soviet Union of Stalin was indeed a black page in human history. Mechanisms must be developed to prevent the possible reoccurrence of political control and manipulation of science. Such control was not limited to the Soviet Union, but could be seen in its deadliest form in Nazi Germany and occasionally surfaces in countries such as the United States when politically ambitious bureaucrats attempt to direct science through selective funding of "politically correct science and thought."

The further internationalization of Soviet genetics will eliminate some of the previous intellectual isolation. This is already taking place with the international co-operation of the Human Genome Program. In particular, there is a realization that small human societies are rapidly disappearing and their genetic heritage must be safeguarded for the generations that follow. The preservation of the molecular past of Siberian and American Indian indigenous populations should be of high priority to the United States, Canadian and Soviet governments. With these ongoing research collaborations between scientists of the Soviet Union and the rest of the world, it is less likely that ideology can once again be imposed upon science.

Some of the bureaucratic structures that have been constructed to monitor and control science in the Soviet Union should be dismantled. The research units should be "streamlined" and made less monolithic. At present, a single director supervises the research of numerous scientists often in fields in which he/she has no competence. It is not unusual, for a director to come from a different field of specialization with his primary qualification being ideological purity! Instead, the researchers should be given more autonomy and the most talented scholars and administrators should be placed in supervisory positions.

At present, little is known about the complex processes of human adaptation. We believe that studies of adaptation should be built upon a solid foundation of molecular genetics. It is essential that the geneticists of the Soviet Union focus more on the normal molecular genetic variation as a basis for understanding how evolution occurs in human populations. The past intellectual isolationism of the Soviet Union can only be overcome by collaboration between their scientists and researchers from other countries. It is with this collaborative spirit that we have written this article. Its purpose is not merely to criticize, but through international co-operation to further the development of science in both the former Soviet Union and the western nations. We can learn from one another!

*
ACKNOWLEDGMENTS

This research was supported in part by a grant from the National Science Foundation BSR-99101571. The idea for the title of this chapter came from the title of an article by Derek F. Roberts, entitled "Physical Anthropology in the United States: The Malaise and its Cure." This article appeared in the *American Journal of Physical Anthropology* 25: 165-168. We thank Professor Roberts for his inspiration.

REFERENCES

Alexseyeva, T.I. 1984. *Anthropological Surveys in Tuva.* Moscow: Nauka. (In Russian).

_____. 1986. *Adaptive Processes in Human Populations.* Moscow: Nauka. (In Russian).

_____. 1989. Regularities in spatial variability of morphological and physiological traits and geographic variation of the "norm" in humans. In: *Anthropology for Medicine,* ed. T.I. Alexeyeva, Moscow: Nauka, pp. 17–36. (In Russian).

Allison, A.C. 1954. The distribution of the sicklecell trait in East Africa and elsewhere and its apparent relationship to the incidence of subtertian malaria. *Trans. R. Soc. Trop. Med. Hyg.* 48: 312318

_____. 1961. Genetic factors in resistance to malaria. *Ann. N.Y. Acad. Sci.* 91: 710–729.

Altukhov, Y.P. 1989. *Genetic Processes in Populations.* Moscow:Nauka. (In Russian).

Cavalli-Sforza, L.L. and W.F. Bodmer. 1971. *The Genetics of Human Populations.* S.F.: Freeman Press.

Chakraborty, R., R.E. Ferrell, S.A. Barton, and W.J. Schull. 1986. Genetic polymorphism and fertility parameters in the Aymara of Chile and Bolivia. *Anns. Hum. Genet.* 50: 69 82.

Chetveryakov, S.S. 1927. About certian features of evolutionary process from the standpoint of modern genetics. Cited from *Classics of Soviet Genetics,* ed. P.M. Zhukovsky. Leningad: Nauka, 1968, pp. 133–170. (In Russian).

Comuzzie, A.G. and M.H. Crawford. 1990. Biochemical heterozygosity and morphological variability: Interpopulational versus intrapopulational analyses. *Human Biology* 62: 101–112.

Corrigan, C. and S. Sega. 1950. The incidence of congenital dislocation of the hip at Island Lake, Manitoba. *Canadian Med. Assoc. J.* 62: 535–540.

Crawford, M.H. 1987. Origin and manintenance of genetic variation in Black carib populations of St. Vincent and Central America. In: *Genetic Variation and its Maintenance,* eds. D.F. Roberts and G. DiStefano, Cambridge: Cambridge Univ. Press, pp. 157180.

_____. and A.G. Comuzzie. 1989. Genetic and morphological variation in the Black Carib populations of St. Vincent and Livingston, Guatemala. *Collegium Antropologicum* 13: 51–61.

_____. and E. Goldstein. 1975. Demographic structure of an urban, ethnic isolate: Polish Hill, Pittsburgh. *Amer. J. Phys. Anthrop.* 43: 133–140.

Crow, J.F. 1958. Some possibilities for measuring selection intensities in man. *Human Biology* 61: 776–780.

———. 1989. Update to some possibilities for measuring selection intensities in man. *Human Biology* 61: 776–780.

Dobzhansky, T. 1969. Evolution of mankind in the light of population genetics. *Proc. XII Internat. Congr. Genet.* 3: 281–292.

Dubinin, N.P. 1931. Geneticoautomatic processes and their bearing on the mechanism of organic evolution. *J. Exp. Biol.* 7: 463–479.

Dubos, R. 1965. *Man Adapting* Yale University Press, New Haven.

Dubrova, Y.E., T.M. Karaphet, R.I. Sukernik, and T.V. Goltsova. 1990. Heterozygosity and fertility relationships in Forest Nentsi and the Ngansans. *Genetika* 26: 122–129.

Eriksson, A.W. 1987. Aspects of genetic epidemiology of the North Calotte. Abstracts. 7th International Congress on Circumpolar Health, Umea, Sweden, p. 18.

Ferrell, R.E., R. Chakraborty, H. Gershowitz, W.S. Laughlin, W.J. Schull. 1981. The St. Lawrence Island Eskimos: Genetic variation and genetic distance. *Amer. J. Phys. Anthrop.* 55: 351–358.

Haldane, J.B.S. 1946. The mutation rate of the gene for hemophilia, and its segregation ratios in males and females. *Ann. Eugenics* 13: 262–271.

Karn, M.N. and L.S. Penrose. 1951. Birthweight and gestation time in relation to maternal age, parity and infant survival. *Ann. Eugen.* 15: 206–233.

Kraus, B.S. and J.R. Schwartzmann. 1957. Congenital dislocation of hip among the Fort Apache Indians. *Bone Joint Surg.* 39: 448–449.

Lasker, G.W. 1969. *Human biological adaptability. Science* 166: 1480–1486.

Lerner, I. 1954. *Genetic Homeostasis.* Edinburgh: Oliver and Boyd.

Levine, P., E.M. Katzin and L. Burnham. 1941. Isoimmunization in pregnancy: Its possible bearing on the etiology of Erythroblastosis fetalis. *J. Amer. Med. Assoc.* 116: 825.

Miller, L.H., S.J. Mason, D.F. Clyde, and A.B. McGinniss. 1976. The resistance to Plasmodium vivax in blacks: The Duffy blood group genotype, Fy ab. *New England J. Med.* 295: 302.

Morch, E.T. 1941. *Chondroplastic Dwarfs in Denmark.* Ejnar Munksgaard, Copenhagen.

Motulsky, A.G. 1960. Metabolic polymorphisms and the role of infectious diseases in human evolution. *Human Biology* 32: 28–62.

Neel, J.V. 1989. Human evolution and the founderflush principle. In: *Genetics, Speciation, and the Founder Principle.* eds. V. Giddings, K. Kaneshiro, and W. Anderson, Oxford: Oxford Univ. Press, pp. 299–313.

Posukh, O.L., V.P. Wiebe, and R.I. Sukernik. 1990. Genetic and ecological studies of aboriginal inhabitants of Northeastern Siberia. III. Demographic structure in three Even's populations in Yakutiya. *Genetika*. 26: 1628–1636. (In Russian).

Reid, R.M. 1973. Inbreeding in human populations. In: *Methods and Theories of Anthropological Genetics*. eds. M.H. Crawford and P.L. Workman. Univ. of New Mexico Press: Albuquerque, pp 83–116.

Rychkov, Y.G., I.V. Perevozcikov, V.A. Sheremetyeva, T.V. Volkova and A.G. Baslai. 1969. Population genetics of the indigenous groups of Siberia. Eastern Sayan. *Voprosi Antropologii* 31: 3–32. (In Russian).

————. and V.A. Sheremetyeva. 1972. Population genetics of the peoples of the Northern Pacific Basin in connection with problems of history and adaptation. *Voprosi Antropologii* 42: 3–30. (In Russian).

————. and V.A. Sheremetyeva. 1977. The genetic process in the system of ancient human isolates in North Asia. In: *Population Structure and Human Variation*, ed. G.A. Harrison, Cambridge: Cambridge Univ. Press, pp. 47–108.

Schmitt, L.H., G.A. Harrison, and R.W. Hiorns. 1988. Genetic and morphometric variances in three human populations. Ann. *Hum. Genet*. 52: 145–149.

Solovenchuk, L.L. 1979. Ecologically conditioned changes in gene frequencies for polymorphic systems observed in newcomers to the Northeast of the USSR. *Genetika*. 15: 1775–1783 (In Russian).

————. 1984. Variability of genotypic structure of human populations in extreme environments. In: *Human Inheritance and Environment*. ed. Moscow: Nauka. (In Russian).

Spitsyn, V.A. 1985. Human Biochemical Polymorphism. Anthropological Aspects. University of Moscow Press, Moscow. (In Russian). Spitsyn, V.A. and A.G. Novoradovsky (1989) Major ideas and concepts in human ecology. In: *Anthropology for Medicine*. ed. T. Alexeyeva, Moscow: Nauka. pp. 37–50. (In Russian).

Sukernik, R.I. and M.H. Crawford. 1984. Population genetics and the first peopling of America by humans. *Preroda* 4: 90–99.

————., S.V. Lemza, T.M. Karaphet, and L.P. Osipova. 1981. Reindeer Chukchi and Siberian Eskimos: Studies on blood groups, serum proteins, and red cell enzymes with regard to genetic heterogeneity. *Amer. J. Phys. Anthrop*. 55: 121–128.

————., R.I., L.P. Osipova, and M.S. Schanfield. 1991. Distribution of GM allotypes and deleted IGHG1 haplotypes in the Se'lkups in comparison to the Forest Nentsi of Siberia. *Experimental and Clinical Immunogenetics* (Submitted).

————., V. Vibe, T.M. Karaphet, L.P. Osipova, and G.A. Kirpichnikov 1986. Genetic and ecological studies of aboriginal inhabitants of Northeastern Siberia. II. Polymorphic blood systems in Asiatic

Eskimos. *Genetika* 22: 2369–2380. (In Russian).

Szathmary, E. 1981. Genetic markers in Siberian and North American populations. *Yearbook of Phys. Anthrop.* 24: 37–73.

Vavilov, N.I. 1927. Geographic regularities in the distribution of genes of acculturated plants. Cited from *Classics of Soviet Genetics*, ed. P.M. Zhukovsky. Leningrad: Nauka, 1968, pp. 51–57. (In Russian).

Wright, S. 1922. Coefficients of inbreeding and relationships. *American Naturalist* 56: 330–338.

CHAPTER 5

SOCIOBIOLOGICAL ASPECTS OF HUMAN EVOLUTION
AND COMPARISONS OF SOCIOBIOLOGICAL APPROACHES IN THE USSR AND CZECHOSLOVAKIA

V. LEONOVIČOVÁ

First of all I would like to begin by noting that my paper is based on a survey of the literature of the USSR and Czechoslovakia, when these countries still existed as socialist states. To understand the present situation of science in these countries in the 1990s, it is important to review the past influences.

The next note must be made on the specific situation in the humanities and social sciences in the USSR. One Soviet psychologist referred to these disciplines as the Galapagos. His remark seems very true to me and reflects the situation that existed there very clearly. Strong isolation of social scientists (including psychologists) from the intellectual traditions and scientific community of the West created a unique situation. It is very difficult, for example, to review for the Western reader, the papers and books on the relationship of the biological nature of Man and his social peculiarities that were published in the USSR not only because of the problem in translating them from Russian to English, but also because it is necessary "to translate" them from one style of thought to another very different style, and in some cases this becomes impossible.

During the time of my university studies from mid-1960 to mid-1970 in the USSR at Leningrad, now St. Petersburg, all sciences on Man were under censorship of ideology. Nearly all attempts to see a biological basis for human behaviour or human society became classified as *biologizatorstvo* and became equated to the enemy's ideology. Biologists chose to study the "neutral" (in an ideological sense) subjects because they wanted to avoid the possibility of ideological evaluation. These limitations, among other things, affected the development of the study of human evolution. For example, when the famous book of Ernst Mayr on *Animal Species and Evolution* was translated to Russian (1968), the chapter "Man" was omitted. It was published later in the journal *Priroda* (Nature). Its publication took a great deal of courage by the editor who added some special philosophical comments to justify its publication.

Studies of human evolution in the Soviet Union and the problem of anthropogenesis, were also limited by the official policy that recognized only F. Engels' theory on the role of labour in the transformation of apes into humans. Some very interesting research resulted from trying to verify Engels' theory. This included the work of comparative psychologists and physiologists in the primatological centre in Suchumi (Georgia), such as Vojtonis (1949) Tich (1963) and Roginsky (1948). This tradition was continued in the Koltishi (Pavlov's Institute of Physiology) by Firsov (1977) and co-workers.

Another attempt at verification of the ideas of Engels was the theoretico-philosophical work of Ju. I. Semenov (1962), *How Did Humankind Originate?*. It was a very interesting publication, especially for Soviet readers who were isolated from the literature of the West. Semenov analysed and synthesized the up to date literature in anthropology, paleontology, ethnology and biology. His approach was influenced by the Marxist concept of society, and this of course influenced his understanding and evaluation of the facts and theories he summarized. His main idea was on the uniqueness of human society and learned behaviour (as compared to instinctive behaviour of animals). This view affected his conclusion that human behaviour and human society separate humans from their evolutionary antecedents.

In Czechoslovakia, however, the intellectual environment was quite different. This was due to their rich tradition of anthropological research, their relatively higher level of general interest in cultural pursuits and less disturbing official dogmatic control over the intellectual traditions in science. The influence of communist ideology appeared most strongly after 1968 (in the period referred to as "normalization") and was used as a divisive tool by the *apparat* of the communist party, resulting in personal competition among some scientists.

In Czechoslovakia the traditional connections with European science were severely cut back for 40 years. However, during this time there continued to be plenty of personal, family and friendly connections that were preserved. In addition, during the last two decades, the Czechs attempted to be the mediator between Western and Eastern scientists through the organization of international scientific meetings. Prague became the place of meeting of scientists where East and West could make personal contacts.

This difference in the situation in the former USSR and Czechoslovakia affected the difference in response by scientists to sociobiology.

From the moment of publication of Wilson's *Sociobiology* in 1975, in East Europe his ideas have been differently received in the two spheres — in biology, and in humanities and social sciences. The biologists, especially in the USSR, just did not respond. They remembered very well the repressions against "bourgeois science" and therefore did not like to discuss the problems relating to this subject. This was the study of human behaviour which had been taken over by ideology, and any "heresy" against Marxist dogma was not permitted.

Social scientists and people studying the humanities reacted negatively. This situation reflected in some aspects the controversy in the American scientific community, especially as seen in discussions in which such authors as S.J. Gould and Lewontin, known as sympathizers with Marxism, have seen in sociobiology some parallels with Naziism. The memory of the nazi application of biology to human society was vividly recalled. Nearly every critical response from Marxist social scientists and teachers of Marxism in the universities, was based on indirect sources and was written from the position of official ideology only. These critical papers were not serious and had no value for the development of science. Most of them were published as critiques of bourgeois ideology, not as scientific analyses.

The study of animal behaviour led logically to the study of social behaviour and the evolution of animal society. In this area also the situation in the USSR and Czechoslovakia developed very differently. In Czechoslovakia this direction of research has been very popular and since 1970 there has been a tendency to integrate the efforts of three groups of researchers: zoologists (the ethological section of the Zoological society), comparative psychologists (a section of the Psychological society) and physicians and physiologists (the section of higher nervous activity of the Purkinje medical society). The problems of sociobiology have been discussed at several annual meetings by scientists in association with the Laboratory of Evolutionary Biology in Prague.

The question of the role of behaviour in the process of evolution was one of the theoretical problems being studied at our laboratory. (Leonovičová and Novák, 1987). The founder of our laboratory V.J.A. Novák (1982) has a special interest in the problem of sociobiology. That is why he formulated the "principal of sociogenesis" as a general rule of evolution. According to him the tendency towards unification is a common theme in evolution. The unification of uni-cellular organisms has led to the origin of multi-cellular organisms, and the unification of individuals into society has led to the origin of a new step of evolution which he terms psycho-social. But his ideas on the mechanisms of evolution are not purely neo-darwinistic. He includes in them some Lamarckian elements and therefore could not agree with Wilson's idea of altruism and natural selection that selects the gene but not the organism. Here we can refer to Wilson's statement: "...the central theoretical problem of sociobiology: how can altruism, which by definition reduces personal fitness, possibly evolve by natural selection?" (1975, p.3).

In our laboratory the analysis of the problem of the role of behaviour in evolution has led us to the problem of sociobiology. Behaviour is one of the main forms of adaptation that has determined the origin and existence of animal and human societies. We felt it necessary to modify the theory of evolution, that in its neo-darwinian form could not explain many evolutionary processes, especially those determined by behaviour.

In the USSR the situation was different, mainly because the study of animal behaviour had declined and was not very popular. Comparative psychologists (such as Fabri, 1976) were constrained by the "taboo" of Marxist doctrine and the Pavlovian concept (accepted as the "official truth"), and did not try to see the continuity of the evolution of behaviour of animal and human. In addition, in Russian literature (which was completely censored) the term "society" (*obshestvo* in Russian) had been used only for human society. Animal societies were called "community" (*soobshestvo* in Russian). This situation became especially confused in the context of biology, where this term is usually used in ecology and means a multi-species group of organisms that co-exist in one ecological niche.

Psychologists who studied the nervous mechanisms of behaviour, were influenced by the dogmatic form of the Pavlovian theory of conditional reflexes. All other theories were usually judged to be influenced by bourgeois ideology.

It has been only the zoologists who began to investigate the behaviour of animals by observation in nature (such as the ornithologist E. Panov) who have tried to analyse sociobiology as a scientific discipline. For example, Panov (1983) in his fundamental book *Animal Behaviour and the Ethological Structure of Populations* (in Russian) referred to his own approach as "socioethology." In my opinion his book is a variant of sociobiology that does not include *Homo sapiens* and human society. The last part of the book "The problem of the evolution of biosociality: contemporary state and future perspective" contains an analysis of many concepts of sociobiology, such as a form of selection (individual, kin, group selection), inclusive fitness, and the evolution of altruism. The author divides all investigations of the evolution of biosociality into two groups according to approaches used: (1) sociobiological and (2) ethologo-phylogenetic, such as the famous work of the Russian entomologist, S.I. Malyshev (1966), on the evolution of the social behaviour of Hymenopthera.

Panov recognizes the principal error of sociobiological theory — its foundation on the idea "one gene — one character." That is especially questionable in the case of behaviour. He concludes: "The fact of the existence of sociobiology is very useful. Its hypotheses have stimulated a large number of field investigations that have resulted in rich empirical material. But the theoretical background of sociobiology has led the study of biosociality into a maze of abiological scholasticism" (p. 292). Panov analysed the principal aspects of classic sociobiological theory — forms of selection, altruism, inclusive fitness, and cost-benefit analysis — and he concludes that such approaches are lacking and have been compounded by an absolute disregard of the main problem — the degree of genetic determination of social behaviour (p. 298).

Instead of the classic conception of sociobiology which relates the origin of sociality to the division of labour in the process of reproduction and the origin of altruism, the author prefers to think about the main ways and principles of evolutionary changes in biosociality. This approach needs a comparison of contemporary data on the ethological structure of population with the data on the phylogeny of certain taxa. The study of temporal and geographic variability in the structures of populations of varied species shows the flexibility of socio-demographic systems that are able to react adequately to almost every environmental change. In sociobiology the mechanisms of evolutionary changes in a socio-demographic system show how one can estimate "future success," but the history of the species is not taken into account. Phylogenesis has created the possibility and constraint of future changes, and is more useful for understanding evolutionary mechanisms and tendencies than is cost-benefit analysis.

Now, let me turn briefly to the situation in the USSR that has changed after the revolutionary events in the 1990s. Ideological control and coercion disappeared, but the external control did not affect the inner constraints of personal philosophy and customs that could not be changed so readily. But new approaches to the analysis of sociobiology as a scientific theory have occurred as can be seen in such authors as R. Karpinskaja (1991), Novozenov (1990, 1991) and others.

For example, the book *Biology and the Knowledge of Man* (Karpinskaja, 1989) consists of three parts, each of them written by a group of specialists in various disciplines: philosophers, psychologists, biologists, physicians. The main goal of the publication formulated by its editor is to find a new (for Marxist philosophy) understanding of human nature that includes contemporary biological data, and the necessity to use a new (for Marxist philosophy) "historical approach" in the study of Man. This "new" approach does not separate the evolution of the biological predecessor of Man from the history of human society. She interprets "historicism" as a unification of evolutionary-biological and humanistic approaches.

I think that sociobiology now is a broader discipline that does not have to be limited by the classic problems formulated by its founder E.O. Wilson. All problems concerning the consequences of the origin and evolution of behaviour as a form of adaptation could be seen as a sociobiological issue.

Homo sapiens as a highly specialized species is adapted to existence in human society and behaviour is his main form of adaptation; it is a biological specialization. An understanding of human evolution, including the process of anthropogenesis, cannot be possible without using a sociobiological approach. Sociobiology as a discipline synthesized behaviour into the context of evolutionary theory.

All societies of higher vertebrates have originated on the basis of behaviour, and could appear only after the occurrence of a relatively high level of complexity of the organisms. Behaviour here is defined as all actions directed by organisms toward the external environment in order to change conditions therein, or to change their own situation in relation to the environment.

It seems that there are some principles of biosociology that are invariant for all societies of higher vertebrates, and without that, no society of higher vertebrates (including Man) could exist.

1

A social way of life presupposes the inner structure of a group concept. The origins of the group of organisms are determined by the outside ecological factors only. These aggregations of individuals of the same species originate as a result of variation in environments. The distribution of organisms in certain places correlates with the distribution of the limited resources. Some species do not have any inner structure and are not societies (e.g., concentration of snakes in cold seasons).

Societies of sexually reproduced species, begin with the co-operation of sexual partners in the process of reproduction. This creates the possibility of a future society adapting by natural selection. Step by step complicated group structure creates the basis of animal society.

Finally, there are societies that exist continually. These societies have an inner structure supported by systems of hierarchy and dominance based on the natural differences of individuals, primarily sex and age. Each individual changes his place in the social structure with a change in its age (and social role). Individuals of higher vertebrates, especially the species with higher levels of intelligence have learned to play social roles by imitation. A hierarchical system secures the individual position of each member of society and prevents fighting inside the society. The leader of a group not only subjugates others to him, but at the same time takes them under his protection. A probable explanation for the need of ancient man to believe in an "all-powerful" and "all-knowing" Being has its roots in this peculiarity of all societies of higher vertebrates.

For animals with various degrees of intellectual abilities, the structure of society correlates with these abilities. A consistent evolutionary approach to the investigation of animal behaviour helps in discovering a large variety of cognitive abilities of various species according to their ecology, food strategy, etc. In general, predators have a higher degree of intellectual abilities than herbivores, and a weaker structure of society. For species with a high level of intellectual abilities, and where leadership is situational, the intelligence of an individual could be more useful in acquiring a place in the system of hierarchy (Hinde, 1970 and others).

2

The ability of the members of the group to recognize and to identify their fellow members and to differentiate them from the others of their own species is an important condition for the social way of life.

The social insect can identify the members of its own society by the help of special mechanisms of smell sensation without any personal identification. The individuals with an alien smell are usually banished or killed. Higher vertebrates, in addition to smell identification, have a complex of signals by which they can identify one another "personally." Personal identification is possible in a small society. Societies of higher vertebrates can accept and include aliens and non-relative individuals (sometimes representatives of other species) as members of the society. For example, this ability is developed in sheep-dogs which are raised as puppies among the sheep, and when the puppies develop into adult dogs they defend the sheep as their fellow members of society. In this case the sheep-dogs look after a herd of sheep without Man and successfully defend them from coyotes (Coppingers, 1982). As well, young dogs brought up in human families behave as a member of the family — they maintain the hierarchy, defend others, expect help and defence from other members of the family.

3

The ability of individuals to influence the emotions of one another is a very important peculiarity of social vertebrates and a necessary condition for the existence of society. It is known that the behaviour of all warm blooded vertebrates is regulated by two systems: (i) the cognitive one that helps the animal to take its bearings in its environment and to react adequately to change; and (ii) the effective one that reflects the up-to-date neuro-hormonal-metabolic state of the organism and helps it to do the fast and useful option in its behavioural repertoire. For example, fear can mobilize the organism to run quickly. Mechanisms of emotion have provided the fast reaction and the state of being awake, the energetic background of all activity, including the cognitive one, and are connected with the oldest part of the brain (the thymencephalon).

In societies of birds and mammals the emotional mood can spread very quickly, e.g., when a group of ungulates follow an alpha male, etc. A panic or feeling of anxiety alarms all members of the group and prepares their fast reaction to a dangerous situation. This translation of moods helps to co-ordinate the activity of members of the society and makes them follow the leader. The emotional influence is realized through the complicated system of pheromones, various smell signals, sounds, poses, gestures, etc. Social animals of these groups have acquired the ability to feel discomfort as a consequence of observing suffering of their fellow members and the ability to recognize the state of others, not only for the possibility to co-ordinate their activity, but to feel "together" also.

4

Communication is an absolutely necessary condition for the social way of life. The uniquely human mode of speech and thought in the form of human language (as a system of symbols with high combinational characteristics) has evolved from the communicative system of ape-like predecessors. Transformation of the society of the ape-like antecedent of Man into human society was primarily determined by the origin of human language. This provided the new possibility of the growth of a common pool of experiences and knowledge of the world (common not only for one generation, but for many subsequent ones). This process occurred in conjunction with the evolution of the brain (Lieberman, 1991), and its sociobiological aspect is obvious.

5

The ability to suppress aggression in inter-individual interaction within society is a very necessary condition of social life. The formation of the mechanisms for such types of behaviour were supported by natural selection, because it provided higher survival possibility of the species as a whole. Many behavioural patterns provided the basis for suppression of aggression (as a rule, it is imitation of offspring behavioural patterns that mean the demonstration of submission). Hierarchical structure also supported the suppression of aggression. Many of these patterns are learned by individuals during early ontogenesis. The processes of socialization of offspring of social birds and mammals are as necessary as for human children.

In conclusion I would like to note that sociobiology is one of the biological disciplines absolutely necessary for the study of human evolution and the understanding of human nature. At the same time it is necessary to point out that sociobiology now is not limited just to the problems formulated by its founder E.O. Wilson. Rather, sociobiology can provide new insights into the formulation and research into evolutionary theory.

REFERENCES

Fabri. 1976. *Foundation of Zoopsychology* (in Russian). Moscow, Publ. house of Moscow University.

Firsov, L.A. 1977. *Behaviour of Apes in the Natural Conditions* (in Russian). Nauka, Leningrad. 162.

Hinde, R. 1970. *Animal Behaviour: a Synthesis of Ethology and Comparative Psychology (2nd ed.)*. McGraw Hill, New York.

Karpinskaja, R.S., (ed.). 1989. *Biology in the Knowledge of Man* (in Russian). Nauka, Moscow. 255.

Karpinskaja, R.S. 1991, Marxist thought. In: Mary Maxwell (ed.). *The Sociobiological Imagination*. State University of New York Press, Albany, N.Y., pp. 243-251.

Leonovičová, V. 1983. Motive forces of the evolution of hominids in the direction of sapientization. *Anthropologie*, XXI/I: pp. 27-31.

———. 1985. On the problem of biological basis of the origin of human society. In: Mlíkovsky, Novák (eds.). *Evolution and Morphogenesis*. Academia, Praha, pp. 729-738.

———. 1985. Behaviour and its role in evolution (in Russian). *Journal of General Biology*. Moscow, XIVI, 6: pp. 753-759.

———. 1987. Behaviour, adaptation and evolution. In: V. Pesce Delfino (ed.). *International Symposium Evolutionary Biology*. Bari, Adriatica Editorice, pp. 269-281.

———. 1991. Origin of human mind as a scientific problem, In: J. Piontek (ed.) *The Peculiarity of Man*, Seria Anthropologia nr. 18, UAM, Poznán, pp. 133-153.

———., Novák V.J.A. (eds.). 1987. *Behaviour as One of the Main Factors of Evolution*. Praha, Academia, p. 362.

Lieberman, Ph. 1991. *Uniquely Human. The Evolution of Speech, Thought and Selfless Behaviour*. Harvard University Press, Cambridge, Mass., and London, England. 210.

Malyshev, S.I. 1966. *Hymenopthera, Their Origin and Evolution* (in Russian). Moscow, Publisher House of Academy of Sci. 327.

Mayr, E. 1968. *Species and Evolution*. Moscow, Mir, 593pp. (In Russian).

Novák, V.J.A. 1975. *Insect Hormones*. Chapman and Hall, London, 600 p.

———. 1982. The principles of sociogenesis. *Academia*, Praha. 214.

Novozenov, Ju.I. 1990. Sociobiology and phyletic evolution of Man (in Russian), In: A.V. Jablokov (ed.). *Biology and Contemporary*, pp. 124-138.

————. 1991. *Status, Sex and Evolution of Man* (in Russian). Sverdlovsk, Ural State Univ. Press. 160.

Panov, E. 1982. Contemporary state and the perspectives of development of evolutionary sociobiology (in Russian). *Zoological Journal*, T.LXI, 7.

————. 1983. *Animal Behaviour and Ethological Structure of Populations* (in Russian). Moscow, Nauka. 428.

Roginsky, G.Z. 1948. *Acquired Habits and the Sources of Intellectual Activity of Apes (Chimpanzees)*. Publisher House of Leningrad State Univ.

Semenov, Ju.I. 1962. *How Did Humankind Originate?* (in Russian). Krasnojarsk, Siberia.

Tich N., 1963. *Prehistory of Human Society*. Leningrad State Univ. Press., Leningrad.

Vojtonis N. 1949. *The Prehistory of Intellect*. Moscow-Leningrad, Publ. House of Academy of Sci.

Wilson, E.O. 1975. *Sociobiology: The New Synthesis*. Belknap Press of Harvard Univ., Cambridge, Mass.

CHAPTER 6

HUMAN EVOLUTION IN MICROCOSM

JAROSLAV SLÍPKA

The real world of science is formed by a microcosm of uninterrupted transformations in the process of recognition of the laws of nature. This microcosm of science is constrained not only by the extent of the problems, but also by the time limits. In the case where we have chosen anthropology as our microcosm of science, then one of the branches of anthropology is the history of Man , i.e., the history which has always two faces — a phylogenetical and an ontogenetical one. Both these views are complementary, and there is no sharp border between them; on the contrary, more and more we realize that we cannot study them separately and that phylogeny can only reflect the history of changes in ancestral ontogenies. Contemporary Man should be considered as a result of the phylembryogenetic alterations.

I would like to review the history of such a phylembryogenetic approach in our country and in our universities, being convinced that it is the Central-European region which has been a cradle of evolutionary morphology. Real morphology is grasped not as morphometry only, but as a science of structure and its history. And it was Anthropology — the science of Man — in which the evolutionary morphological methods were most useful.

Czechoslovakia is a small country in Central Europe — a country which represents a heart of this part of Europe. Not only is it a geographical heart, but in the past it was also a cultural centre (in some cases — e.g., in music — also today). Allow me to choose as an example the science of anthropology as an integral part of our culture.

The history of the Science of Man can be divided in three main stages.

1

From Middle Ages to the end of 18th century. The first period of mysticism, astrology, etc. culminated by Thaddaeus Hagecius (1521-1600) who was the "Protomedicus of the Czech Kingdom" and wrote a book *Metoscopy* — the teaching on wrinkles (curved wrinkles = bad morals, horizontal wrinkles = good morals). This was probably the first "anthropological book," published in Latin, in our country at the end of the 16th century. The second part of this period could be dated from 1600, the year in which Jan Jessenius from Jeseni (1566-1621), the President of the Charles University in Prague, performed the first public postmortem examination in Prague and started the interest in anatomy. The peak of this first stage had been reached abroad in Sweden by the father of natural sciences,Carl Linné (1707-1778), and in France by Georges Buffon (1707-1788). Both books — *Systema Naturae* (1735), as well as *Histoire Naturelle Generale et Particulière* (1749) very much influenced our scientist, Sv. Pressl (1791-1849), Professor of Zoology at the Faculty of Medicine in Prague.

2

The second period is characterized as the scientific conception of human beings. It started from the Linnéan and Buffonian time and continues via Purkyně to Darwin. This was the golden age of Czech science, and we shall describe it more thoroughly further. The next stage starts in the Darwinian era and proceeds across Haeckel and our Grégr and Czermak to the first teacher of anthropology, Professor Lubor Niederle, who was recommended to the subject of Anthropology by our great philosopher and first president of Czechoslovakia, Professor Tomáš Garrygue Masaryk. It was 1891, i.e., one hundred years ago, when the history of Czechoslovak scientific anthropology started.

3

The third period comprises the last hundred years and can be characterized as a struggle for an independent state and after the First World War by the effort for building up a democratic society and cultural prosperity connected with foundations of new universities and anthropological departments. But very soon the Second World War broke out and fascism crushed our intelligentsia, closed the universities and executed many patriotic students and teachers. The "Museum of Man" was transformed into "Rasseninstitut" and Anthropology was supposed to serve the perverted philosophy of racism.

After the war, we tried again to pick up the democratic traditions. But only three years afterwards a totalitarian regime came again which was accompanied by a totally ideologized science — science which was obliged to serve the aims, determined in advance. Such science was supposed to adapt itself to the ideology — not contrarywise, i.e., the results must suit the philosophy. If in fascism Anthropology should bring evidence against racial equivalence ("racial anthropology"), then in communism the "class anthropology" was supposed to confirm the "Share of the work in humanizing the ape" (Engels, 1876).

In the situation of ideologized biology which evolved from the unqualified vulgarization of the Marxist concept of science, mainly the Science of Medicine has preserved the possibility to continue in objective cognition and exploration of the actual reality. And it is not a paradox that research in evolutionary biology (especially morphology), which had a deep tradition in our country, has reached a relatively good standard. Even in continuously increased isolation in which it was very difficult to share our results, we have reached some achievements about which I wish to speak later.

We have to take into consideration that science forms one of the most important constituents and factors of the human culture. And cultural development has always been connected with development of the educational system. In this respect my country has very rich traditions. It was Czechoslovakia that, through its first University, helped to light the flame of European culture and civilization. It was very soon after that, that the foundation of Universities in northern Italy, Paris, Oxford and Cambridge began. And it was our Czech king Charles IV who — himself one of the greatest European intellectuals — created in Prague in 1348 a University, named later in his honour. In his post as a Roman emperor, he then founded Prague University, and afterwards created another nine new universities in various European countries.

One of the first rectors (presidents) of Charles University was a man known throughout the whole world — the famous religious reformer Jan Hus. One of the next presidents was Jan Jessenius, one of the founders of modern anatomy, whom we mentioned above. But one of the greatest biomedical personalities of this school was Jan E. Purkyně — the real founder of Anthropology in our country. Regardless of the fact that Purkyně was the first to formulate the cellular theory in 1837 (two years before Schwann!), he was involved in many branches of human physiology and morphology (Purkyně's cells in the cerebellum, Purkyně's fibres in the myocardium, etc.).

Purkyně, in his speech given on the occasion of the opening of the Physiology Department at the Charles University on June 18, 1851 (just 140 years before), proposed:

> to place anthropology in the vestibule of the science of physiology so that the student gets acquainted with human beings at first. Together with anthropology, the science of botany and zoology should proceed. The subject branches — it describes at first anatomically the form in highly developed organism and it ends by microscopic investigations. The whole can be called morphology. It

includes the anatomy of men and animals — so called comparative anatomy as well as plant anatomy. The microscopic investigation discovers the elements of organisms and leads to histology, i.e., the science of tissues. The only description of the patterns is called morphography — if it compares and explains the significance of the organic forms, it is morphognomy. The development of patterns is studied in morphogeny (of the whole body = embryogeny, of the organs = organogeny, of their primary elements = histogeny) by the way of description, comparison and explanation. After finishing this work, the investigator should return to the morphology of highly evolved organism.

Purkyně in his article "Man and Nature" formulated his evolutionary views two years later (1853):

Nature has created through the gradually complicated forms — from the simplest to the most perfect ones — the whole line of animals, reaching the man, who became the crown of all creatures."

And his views on the problem of recapitulation formulated in the same article as follows:

The fruit of the womb of its mother repeats all the patterns, formed in preceding ages to reach to the present time.

From this we can see that Purkyne was surely prepared — six years before the *Origin of species* was published — to accept the biggest discovery of the 19th century — the evolutionary thoughts of Charles Darwin.

One of his students, Eduard Grégr, was a great propagator of Darwinism. He wrote in 1856 (three years before Darwin) in his paper, "Man in relation to animals,"

Nature has continued, from the first infusorian to the giant animals of the antediluvian era, in formation of more perfect creations till they finally reached the stage of a man....

but he went even further and asked if Man is the final stage of evolution and does not represent a first link of a new line of intelligent creatures?

At the same time another great scientist lived in Brno, the promoter of genetics and discoverer of the genetic laws, Gregor Mendel (1822-1884), whose publication *Versuche über Pflanzenhybriden* (Brno, 1865) remained unnoticed, even though it signified a revolutionary discovery in biology. The exceptional nature of this discovery was ignored because of the immense interest created by Darwinian descendence theory. It took 40 years until the significance of the discovery of a Moravian prelate was properly appreciated, and his laws of genetics have been proved to be valid generally for all organisms.

It is necessary to add to these two giant personalities of our biology in the 19th century — Purkyně and Mendel — a third one, whose merits in anthropology have been world-known. It was Dr. Aleš Hrdlička (1869-1943). Coincidentally, we commemorated a centenary of Purkyně's death in 1969 and at the same time a centenary of Hrdlička's birth. When he was 13 years old, this Czech boy emigrated with his parents and six brothers and sisters to New York. He studied Medicine, but afterwards became an Anthropologist and finally the Director of the Physical Anthropology Section of the National Museum of the U.S.A., in Washington, which developed into the biggest institution of this kind in the world. He also founded the largest anthropological journal, *American Journal of Physical Anthropology*.

Dr. Hrdlička visited his native country in 1922, and he outlined a program of Czechoslovak Anthropology and helped (even materially) to create an Anthropological museum in Prague as well as a Czechoslovak journal, *Antropologie*. He was awarded a degree of honorary Doctor of our Charles University in Prague. He is still very popular in our country because he has never forgotten his origin and he was very helpful in building up the Czechoslovak state after the First World War. As such, he was an honorary consul of our state in Washington. Up to now, our Anthropological Society awards to distinguished scientists the "Hrdlička's medal," with an inscription of his scientific creed: "All mankind is of one origin." Every year in his birthtown, Humpolec, scientific competitions are organized for young anthropologists. The best submitted paper is awarded a Hrdlička's prize (I am very proud that this prize has been given to one of my students for his paper on teeth development).

Hrdlička's ideas have been further developed by his friend and Physician-Anthropologist who became Professor in the first Department of Anthropology in the Austrian-Hungarian monarchy. Professor Jindřich Matiegka (1862-1941) was a real founder of the scientific approach to Anthropology in our country. His main interest was the prehistoric, craniological and osteological research of our population. His best known book is *Homo predmostensis*.

Matiegka's successor was again a physician, Professor Jiří Malý (1899-1950), who collaborated with Hrdlička's anthropological investigation in Alaska as well as in research of the influence of the American environment on the physical properties of Czechoslovak immigrants. At home he studied the remains of our historical personalities and the skeletons of diluvial humans from Dolní Věstonice. I am very happy to be considered as one of his students. He died very early and his successors in his department and other institutions (e.g., National Museum) were V. Fetter, Ch. Troníček, M. Prokopec, K. Hainiš, L. Malá, E. Vlček M. Stloukal, E. Strouhal, V. Vančata and others.

Together with the development of Anthropology at the Faculties of Natural Sciences, the research in comparative anatomy and evolutionary morphology has been concentrated in departments of Zoology. First of all, it was a great scientist, Professor Antonín Frič (1832-1913), who laid down the foundations of scientific zoology in our University and in the National Museum in Prague. Even though he was mostly engaged in faunal studies, he was also very much interested in Palaentology and Microscopy. His student and successor, Professor František Vejdovský (1849-1939), from 1892 Professor of Zoology, Comparative Anatomy and Embryology in our university, was even more important for furthering the development of evolutionary theory. He was an honorary doctor of Cambridge University, and was very much engaged in Cytology (he made the discovery of the centrosome) and Embryology and was a great propagator of Darwinism. Some of his papers were written in co-operation with his successor, Professor Antonín Mrázek (1868-1924), who published an original book, *The Doctrine of Evolution* (1907).

After him came Professor Julius Komárek (1892-1955), author of *General Zoology*. His colleague, Professor Jaroslav Storkán (1890-1942), was executed by the Nazis, together with many Czech patriots, during the Second World War. At this time our Universities were closed by the Germans and during their totalitarian regime, there was a great isolation in the Sciences. Old specialists (Professors Bartoś, Jírovec, Kramář, etc.) were gradually replaced by new ones. There has been progress in experimental embryology, which was introduced by Professor F. Sládeček (1916-) and Professor A. Romanovský (1932-) who studied mostly the problems of cell differentiation and regeneration.

It is necessary to include in Departments with an evolutionary background the Department of Paleontology of Charles University under the heading of Professor Josef Augusta (1903-1968), the author of a series of evolutionary publications (e.g., *Primeval Man*, 1960), and his successor, Professor Zdeněk Spinar (1916-), the author of *Life before Man*, which has been translated into 13 languages. We should not forget the laboratories in Moravia and Slovakia — they have been mostly established after the First World War. In Brno an excellent museum, "Anthropos," has been founded, and around this institution and in the University laboratories worked some very active scientists like Professors Absolon, Zlábek, Jelínek, Dokládal, Novotný, černý and others. Similarly, in Bratislava there were many very active Czech scientists, and gradually a good anthropological basis for research came to be developed (Professors Valšík, Stanek, Pošpísil, etc.).

It is impossible to erect an exact boundary between the Anthropological Departments and Departments of Medical basic sciences. Medicine should be considered as an applied anthropological science and both have developed mutually. In Czech anatomy, some brilliant stars shone which should not be forgotten in Anthropology. It was at first Professor Jan Janošík (1856-1927) who published a series of general embryological works on gonads and cleavage of ova, on germ layers, and also a textbook of anatomy.

His successor, the distinguished Professor Karel Weigner (1874-1937), after writing excellent textbooks of topographic anatomy, turned his attention purposefully to Physical Anthropology. He published the results of his research of the weight of the human brain and its relation to intelligence, and he was an editor of a publication, *Equivalence of the European Races* (1934), in which scientific arguments were brought against Nazi-German racism. His successor was Professor Ladislav Borovanský (1897-1971), distinguished scientist and teacher of anatomy in Charles University who worked mostly in growth anatomy. His study of growth of the body, and the process of ossification, which was documented by x-ray examinations, belongs even today to the classical anthropological literature. All his studies were supported by a large collection of skeletal material, especially in craniological publications, e.g., craniometry of the newborn skulls, sexual differences of skulls, etc.

The successor of Srdínko in the Department of Embryology was Professor Zdeněk Frankenberger (1892-1966). He was an exceptionally well educated embryologist with a wide biological scope. All of his papers were of evolutionary direction, and from the pure anthropological point of view, I shall mention his two publications on the Anthropology of Slovakian people and his book *On the Origin of Man*. Frankenberger was a man of high moral qualities and a convinced advocate of democracy (his brother was executed by Nazis). His publication *Comparative Embryology and Phylogenesis* (1956) could be printed only with an attached leaflet with a comment from the editor who condemned its non-Marxist point of view.

In connection with the personality of Frankenberger, we have to mention also the name of his colleague and friend, Professor Jan Florian (1897-1942), a prominent Embryologist from Brno, the author of papers on the development of the youngest human embryos and co-author (with Frankenberger) of our largest textbook of embryology. This distinguished man was executed by the Nazis! The Department of Anatomy in Brno was headed by the well-known Professor Karel Zlábek (1902-1983) who was persecuted at this time, and then by a skilled embryologist Professor Karel Mazanec (1922-1967) who continued in the studies of Professor Florian on human blastogenesis.

The intensive development of the basic medical Sciences and Anthropology was interrupted by the closure of the Czech universities by the Nazis during the war. After the war, the whole country was in a very difficult state. The Universities were closed for six years — afterwards, there was no young generation of scientists and physicians. One of the solutions was to establish new Colleges. In such a way a Faculty of Medicine has been founded in Pilsen. The chair of Anatomy was taken by Professor Jaroslav Kos (1917-) who was interested in osteology and the structure of synovial membrane. His colleagues who continue in a similar line are Professors J. Heřt (persecuted by the regime), J. Hladíková and P. Fiala. Three of his assistants left the country during the totalitarian regime.

After World War II, Professor Otto Slabý (1913-) from Prague's Frankenberger's Department assumed the chair of the Department of Histology and Embryology. His two assistants, left the country for political reasons. At first very primitive conditions existed in laboratories, even though we got some basal equipment from UNRA (an American aid organization). The lack of equipment has constrained our scientific possibilities for research.

Afterward, we more or less returned to the problems of Purkyně's school and continued in the rich traditions of Czechoslovak biology. We developed the evolutionary approach to the embryological problems and we tried to form a causal, dynamic and historical concept of morphology. For us, morphology means a science of developmental causes of structures and forms, i.e., of their evolutionary history.

Professor Slabý published a large series of studies on the development of the limbs in the whole evolutionary line of vertebrates. He reached many important conclusions, mostly on the morphogenesis of carpal elements which allowed him to generalize his findings in his monograph on the laws of evolution. His studies on the organ of Jacobson, the nasal capsule and its septum and on the segmental origin of the skull are classic. Again and again we have returned to the problem of head morphology, i.e., the craniological theme, introduced in biology by the famous German poet J.W. Goethe, and in craniology by Purkyně.

Like my teachers, I also (Slípka, 1926-) tried to prove that the 200-year-old disputes of morphologists on the antagonistic theories of segmental and non-segmental head origin are only variations on the same developmental theme. We have demonstrated the signs of segmentation of the cephalic part of the notochord in the early development of representatives of all vertebrate groups. Segmentation can be proven also in seemingly non-segmented prechordal part of the head in the earliest stages even at the lower evolutionary stages of vertebrates (e.g., sharks).

The problem of head segmentation is closely related to the questions of branchiomery and the derivatives of the branchial region. These problems have formed the second direction of our research activities — the phylembryology of the immune system. We have been mostly interested in the development of the tonsils from an evolutionary morphological point of view, and we studied some of their features that are similar to the thymus.

The development of lymph organs in germ-free pigs, as well as the comparative studies of age and accidental involution of thymus and tonsils are the theme of our many publications and lectures.

In our Department we have resumed the pursuit of traditional problems, approached at the time of Purkyně and his colleagues. As was mentioned above, one of his assistants (Czermak) was very much interested in mummies. In co-operation with the Czechoslovak Institute of Egyptology, we established a small laboratory in 1973 for paleohistology. Dr. Němečková has developed some special methods for histological examinations of Egyptian mummies. We have studied hundreds of specimens of various tissues (skin, vessels, nerves, cartilage, bone, hairs, etc.) not only of human, but also of various animals and plants. In addition to these mummified tissues, we examined microscopically some remnants of our historical personalities.

Another Purkyně assistant (Räuschel) worked intensively on the structural problems of blood vessels. In our Department, we have been interested in the structural and developmental problems of vessels (blood and lymphatic) from the evolutionary morphological point of view. Associate Professor Kočová from our Department came to the conclusion that evolution as well as the development of the venous system is directed from the superficial to the deep veins. They plan a determining role in the process of human adaptation leading to bipedalism and in the problems of blood outflow from lower extremities. She has found differences in the relationships between the superficial and deep venous system, as well as in the course, size and structure of veins in mammals with different locomotion and forms of their extremities.

But let me speak in more detail on one of our important research activities. During the last 10 years we have studied human malformations in spontaneous abortions. The reason has been a very practical one. There is an increase in spontaneous abortions (and at the same time a decrease in deliveries), and that is why we formed a team of embryologist, geneticist and pathologist to examine every miscarried case and to help the gynecologist to improve the prognosis of the next gravidity. At the same time, we asked ourselves if these abortions can be considered the result of natural selection during ontogeny, if they can be compared with Darwinian selection, and if their malformations have some evolutionary background.

We embryologists study both the causal and formal pathogenesis and consider malformations as a deviation of the originally normal development, that is to say, an interference of genetic and epigenetic strategies. And we forget very often that the history of a structure (also abnormal) also has a second face — the phylogenetic one. We have to take into consideration that the same environmental pressures can influence not only the process of ontogeny but also the general process of phylogeny.

Evolution of living matter did not proceed gradually. The relative calm periods were characterized by continuous accumulations of new genetic information which evolved by sexual combinations of genomes, but also by positive or negative mutations, capable of potential expression under suitable conditions (preadaptation). These periods alternated with the relatively short periods of drastic changes of the environment, after which the previous conditions have never been repeated.

We have to accept that mutations appeared also among differentiated body cells, i.e., the target of the environmental pressure was the phenotype. There is still discussion on how the phenotype integrates into the hereditary DNA. We cannot exclude the possibility of some germline flow of genetic information, i.e., the inheritance of acquired character as is recently discussed by Steele (1981).

Most of the genetic variations are random and if they are unfit for survival, they are eliminated by the environment. Those species that survive have taken advantage of new genetic information for new adaptive radiation, and the transitional forms of species have become extinct. Their places were invaded by new biological species equipped by a higher degree of their internal morpho-functional organization and hierarchization. These were successful species, forms, which have succeeded in passing the sieve of selection, i.e., in solving the problem of adaptation and survival in a new environment. New species and even new taxa originated most probably just at the time of these "revolutions of evolution."

In all these key periods of revolutionary changes of environment, natural selection had to construct more and more prosperous forms, i.e., more perfect than the previous ones. To fulfil the new functions, new structures evolved, but for their construction only the former existing building material was used. If the evolutionary changes arose as a result of dissociation of inductive influences, we can expect that the remnants of the original arrangement shall persist and that they will be still present as regulatory genes, not spent in the old way.

Changes of the environment are not only climatic but they involve also changes of other physico-chemical conditions such as in antigens, because the higher organisms evolved in co-evolution with antigens. In order to successfully overcome all of these alterations, but mostly biological changes of antigens, they had to build more and more perfect regulatory systems which reached their peak in the immuno-neuro-endocrine system. The regulatory systems represent organs and vectors of defence, which enable the organism to defend against invading pathogens. This function is but a small one if we consider that the body of a modern human consists of approximately 10^{13} (ten thousand billions) of cells and on these cells fall about 10^{14} (one hundred thousand billions) of microbes, i.e., ten microbes to one cell.

Taking all these changes into consideration, we have to adapt our concept of ontogeny as a sum of phyletic changes of previous ontogenies, i.e., the continual sequence of morphogenetical events in the embryo follows the continual sequences of morphogenetical stages built up by previous series of embryos. Ontogeny recapitulates the phylogenic changes which are imprinted in it and which have been determined by the direct activity of the environment. That means the environment not only interferes with ontogeny, but it is also the same environment which represents the factors of natural selection. From all these results, we see that ontogeny and morphogenesis reflect the evolutionary history of the individual and that malformations may not be devoid of evolutionary information.

There is no question about the evolutionary channels in homeotic mutants which we know mostly in Arthropods. The situation in Vertebrates becomes, of course, more complicated. There is also no question about the origin of several malformations with an evolutionary message — supernumerary nipples in humans, tail persistence, hypertrichosis, interdigital webbing in Man, etc. But what about the majority of malformations in which the phylembryological history remains more or less mysterious?

Developmental defects have often been considered as playing a significant role in phylogeny, and the "hopeful monsters" of Goldschmidt should represent the sudden mutants, through which the process of macroevolution could be realized. We don't want to deal with this problem here, and instead, we would like to speak about quite contradictory monsters, i.e., about "hopeless monsters, monsters which have, at first view, nothing to do with evolution. Nevertheless, we shall try to trace the relations and constraints of evolution in the process of pathogenesis of these ontogenetical deviations to transform the hopeless monsters into the helpful ones.

To study this problem, we have examined (in co-operation with the Department of Obstetrics), the human embryonic and fetal material from spontaneous abortions. From approximately 3,000 deliveries (in a year), 400 are lost — a total of 10-15%! But these are only the waste we can observe. Many more embryos are surely miscarried in the first stages of development, and it has been estimated that about 70% of all conceptuses are lost.

We are inclined to speak about the "natural selection" responsible for these abortions. But can we really compare this elimination sieve of non-acceptable ontogenies with the direct action of the environment in the course of evolution? We have to admit that in certain cases the two mechanisms are similar. In many cases of abortuses, the causes lie outside the embryo, i.e., in the lethal activity of maternal factors. These are factors (uterine, incompetence, bleeding or infection during pregnancy) which usually do not cause the abnormalities.

We have chosen for our study a set of anomalies which were either aborted in the embryonic or fetal period, or they were born with heavy defects and died, or lived as handicapped children, treated by surgeons. All of them are malformations of the human face, especially its leading structure, the nose. We have, of course, to keep in mind that the formation of a flattened human face has been a very complex process, based primarily on the neotenic manner of the whole head development.

The nasal malformations could be divided into "absence of the nose" and then "defect in its position." And there are only a few possibilities — either the shifting and merging of both nasal anlages together forming one naris only, or the opposite, i.e., separation of the primordia, which results in cleft nose of various degree. In the frame of these processes also a shifting of the olfactory placodes dorsally can be classified — in this case instead of the nasal wings, a sort of proboscis develops. The changes in nasal morphogenesis are accompanied by the changes in eye position and telencephalic configuration.

The general inductor in this preotic region is represented by the prechordal mesoderm. It has its own history in the anterior arch- archenteron and it spreads in the head mesenchyme (with contribution of neural crest) and forms the branchial structures and induces the ectodermal derivatives — epidermis and prosencephalon. We can distinguish at least four grades of inductors of morphogenetic systems, and this cascade can be the target of the external environment which can interfere in some interactions and consequently damage the destination structures.

Even a tiny change of an exact timing and exact location or shifting of induction can affect the genome or epigenetic interactions to evoke substantial changes in interactions during the state of dissociation. This dislocation influences also the successive integration so that an abnormal structure, like a proboscis, develops. We consider the integration as one of the most important developmental events (which deserves to be more and properly studied) and which could throw more light on the development of malformations. We can suppose that these mechanisms which are effective in a very short time resemble the mechanisms of evolution.

Regarding the changes of localization of the olfactory placodes, we tried to designate the position of the olfactory organ in the representatives of various vertebrate classes by injection of a contrast material. In sharks, the olfactory organ appears in the form of paired nasal sacs, separated from the mouth cavity. In bony fishes, the blind nostrils are shifted dorsally, but in Amphibians they join the mouth through passages (the telencephalon becomes divided into two hemispheres). In Reptiles, the olfactory part becomes differentiated from the respiratory one (the organ of Jacobson appeared). In birds, the nostrils are situated also dorsally and smell develops very weakly.

In mammals and in Man, the olfactory organ develops in a well known manner, i.e., it recapitulates at first the blind tube and later it joins through the choanae with the mouth cavity. We can conclude that there is a tendency to shift the nostrils during evolution from the ventral position (as we know them in sharks) to the dorsal side in a different distance from the margin of the upper jaw. But it is difficult to generalize because there are many varieties inside one group of animals. We can confirm this in the group of Primates which can be divided according to the nasal form in Platyrrhina (with a broad nasal septum) and Catarrhina (narrow septum) and inside this group a great variability of nasal form exists. Everybody knows the extreme development of the nasal form in elephants and in some hoofed animals, like the tapir or even in seals.

These different forms and locations of the nose that evolved in the course of millions of years in various animals cannot be related to the similar locations and forms we have seen in malformed human fetuses. These similarities surely do not prove any phylogenetical relationships between, say, elephant and Man (only because of the proboscis). But they prove that the production of the described anomalous novelties of structure, the same as morphological novelties in evolution, should not be random. The formal genesis of malformations (as well as evolutionary novelties) can only follow a way which has been constrained by a chain of preceding ontogenies — bearers of evolutionary information.

SUMMARY

The author has shown in an historical outline that Czechoslovakia has a long cultural tradition (Charles University since 1348). Among his compatriots he discusses the world-known scientists of the 19th century who contributed fundamentally to the development of biology and anthropology — J.E. Purkyně, G. Mendel, A. Hrdlička, as well as scientists less known like Matiegka, Weigner, Borovanský, Frankenberger and others. He emphasizes that the last one hundred years have been marked in their first half by the struggle of Czech and Slovak nations for an independent state, for a democratic character and finally against a racist nazism. The second half of this century was characterized by another totalitarian regime with a science under the rigid control of ideology. This situation together with an isolation from the rest of the world has caused a substantial delay in some directions of research, mostly in experimental methods. But on the other side, Czechoslovaks have reached quite a reasonable standard in the solution of evolutionary morphological problems.

The author discusses some data of his research on malformations in spontaneous abortions. He is persuaded that an external interference in the genetical material and genetically controlled inductive processes must cause changes of only these structures and functions which have been collected through evolutionary ages, i.e., it can awake the "dreaming genes" whose products did not pass the sieve of natural selection. The embryos and their malformations must be considered as bearers of evolutionary information.

REFERENCES

Absolon, K. 1938. *Exploration of the Diluvial Station of the Mammoth-hunters in Dolní Vestoniče in Pavlovice-hills in Moravia*. Brno, (in Czech).

Augusta, J. 1960. *Primeval Man*. Orbis, Praha.

Borovanský, L. 1936. *Sexual Differences on the Human Skull*, CSAV, Praha, 1936.

Cerný, M. 1971. Geschlechtsbestimmung nach dem postkranialen Skelett. *Nat. Museum*, Praha, 46-62.

Czermak, J. 1852. Beschreibung und mikroskopische Untersuchung zweier ägyptischer Mumien. *Sitzb. Kais. Akad, Wiss. Math.* Nat. 9, 427-469.

Dokládal, M. 1976. *Human Growth and Physical Development*. Univ. JEP, Brno.

————., J. Brozek. 1961. Physical anthropology in Czechoslovakia. Recent developments. *Current Anthropol.* 2, 455-478.

Fetter V., S. Titlbachová, Ch. Troníček. 1956. The evolution of characteristics of adult population in Czech lands during the last 60 years and the basic anthropological norms. *Univ. Carol. Biol.* 2, 209-232 (in Czech).

Florian J., Z. Frankenberger. 1936. *Embryologie*, Melantrich, Praha (in Czech).

Frankenberger Z. 1941. *On the Origin of Man*. Praha, Melantrich (in Czech).

Grégr, E. 1856. *Man in relation to animals*. Ziva, Praha 6 (in Czech).

————. 1858. On human skulls generally and the Slavonic ones especially. *Ziva*, Praha, 8, 223-242.

Hagecius T.H. 1584. *Aphorismorum Metoscopicum Libellus Unus*. Francofurti.

Hainiš, K. 1963. Morphological changes on the skeletons of the Czech population during the last millennium. *Acta Congr. Anthrop. Mikulov*, Brno, 89-96.

Hrdlička, A. 1924. On the origin and evolution of man and the future of mankind. *Koči*, Praha (in Czech).

Jelínek J. 1952. Beitrag zur phylogenetischen Stellung der mitteleuropäischer Neandertalerfunde. *Acta Mus. Moraviae*, 37, 249-260.

John, H.J. 1959. *Jan Evangelista Purkyně - Czech Scientist and Patriot*. Amer. Phil. Soc., Philadelphia.

Kočová, J. 1989. Structure and development of the veins and lymphatics. In: *Phlébologie 89*, A. Davy, R. Steiner eds., J. Libbey Eurotext Ltd., 7-9.

Kos, J., J. Wolf. 1972. Les menisques intervertébraux et leur rôle possible dans les blocages vertebraux. *Ann. Med. Phys.* 15, 203-218.

Malá, L., M. Prokopec, Ch.Troníček. 1956. *On the Traces of Human Evolution.* Orbis, Praha (in Czech).

Malý, J. 1928. Jean Ev. Purkyne anthropologue. *Anthropologie,* 6, 11-17.

————. 1933. Czechoslovaks abroad, esp. in North. America. In: *Ceskoslovenská vlastiveda* 2, 260-269 (in Czech).

Matiegka, J. 1921. The testing of physical efficiency. *Amer. J. Phys. Anthrop.,* 4, 223-230.

————. 1931. History of the science of Man in Czech countries. In: J. Vinklár (ed.): *Vývoj české přírodovedy.* Přír. klub, Praha, 59-63 (in Czech).

————. 1933. Physical anthropology of the population of Czechoslovakia. In: *Ceskoslovenská vlastivěda* 2, 115-254 (in Czech).

Mendel G. 1865. Versuche über Pflanzenhybriden. *Verh. Naturf. Vereines,* Brünn, 3-47.

Němečková, A. 1991. Historical analysis of the mummified tissues from Saqqara (Egypt). *Plzen Med. Rep.,* 64, 39-41 (in Czech).

Niederle, L. 1889. *An Outline of the History of Anthropology.* Praha (in Czech).

————. 1929. Historical records of the type of the ancient Slavs. *Anthropologie,* Praha, 7.

Novotný, V. 1968. Über die Bedeutung des Os pubis für die Geschlechts-unterschiede am Becken der Makaken. *Anthropologie,* 6, 7-17.

Pospísil , M. 1959. Morphologie der Hand der Lausitzer Serben. *Acta Bratisl.* 3, 347-360.

Prokopec, M., J. Suchy, S. Titlbachová. 1953. *Textbook of Anthropological Methods.* Praha (in Czech).

Purkyně, J. Ev. 1918. *Opera omnia.* Vol. I, Praha.

————. 1937. *Opera omnia.* Vol. II, Praha.

————. 1939. *Opera omnia.* Vol. III, Praha.

————. 1941. *Opera omnia.* Vol. IV, Praha.

————. 1951. *Opera omnia.* Vol V, Praha.

————. 1954. *Opera omnia.* Vol.VI, Praha.

————. 1958. *Opera omnia.* Vol. VII, Praha.

————. 1960. *Opera omnia.* Vol. VIII, Praha.

————. 1965. *Opera omnia.* Vol. IX, Praha.

Raeuschel, F. 1836. De arteriarum et venarum structura. Dissert. Univ. Vratislaviae (in Latin).

Skerlj, B., J. Brožek. 1952. Jindřich Matiegka and the development of Czech physical anthropology. *Amer. J. Phys. Anthropol.* 10, 515-519.

Slabý, O. 1952. Contribution to the problem of the segmentation of the vertebrate head. *Folia Biol.* 32, 306-316 (in Czech).

————. 1960. Die frühe Morphogenesis der Nasenkapsel beim Menschen. *Acta Anat.* 42, 102-175.

————. 1968. Wege und Gesetzmässigkeiten der Evolution in Bezug auf die phylogenetische Entwicklung der Extremitäten. *Acta Univ. Carol. Med.* 35, 1-150.

Slípka, J. 1981. Evolutionary morphology of inconstant structures of epipharynx. *Plzeň Med. Rep.* 44, 5-41 (in Czech).

————. 1983. Neotenic features in development of the human face. In: V.J.A. Novák (ed): *General Question of Evolution.* CSAV, Praha, 385-392.

————. 1988. Palatine tonsils - their evolution and ontogeny. *Acta Otolaryng. Suppl.* 454, 18-22.

————., O. Slaby. 1989. Teratology and evolution. *Plzeň Med. Rep.* 59, 41-46.

Spinar, Z. 1973. *Leben in der Urzeit,* Urania, Leipzig.

Steele, E.J. 1981. *Somatic Selection and Adaptive Evolution.* Univ. Chicago Press, Chicago.

Stloukal, M., H. Hanáková. 1966. Anthropologie der Slaven aus dem Gräber-feld in Nové Zámky. *Slovenská archeol.* 14, 167-204.

Strouhal, E., L. Vyhnánek. 1979. *Egyptian Mummies in Czechoslovak Collections.* Sborn.Nár. Mus., Praha 35B.

Studnička, F.K. 1927. Joh. Ev. Purkynjes und seiner Schule Verdienste um die Entdeckung tierischer Zellen und die Aufstellung der "Zellen" Theorie. *Práce Mor. Přír. Spol.,* Brno 4, 97-169.

Vančata, V., V. Přívratský, D. Hellerová, K. Zemek. 1981. Biological prerequisites of hominization. *Anthropologie,* Brno, 14, 237-242.

Vlček, E. 1969. *Neandertaler der Tschechoslovakei.* Academia, Praha.

————. 1986. Purkyne's introduction and heritage to anthropology. In: E. Trávníčková (ed.): Jan Evangelista Purkyně, Avicenum, Praha, 348-361 (in Czech).

Weigner, K. 1934. *The Equivalence of the European Races.* Akademie, Praha (in Czech).

————., J. Bělehrádek. 1936. *The Science of Man.* Prof. nakl., Praha (in Czech).

Zlábek, K. 1929. Sur l'influence du mécanisme du l'articulation temporomaxillaire sur la forme du digastrique de l'Homme et des singes anthropoides. *Bull. Int. Acad. Sci. de Bohème* 39, 1-38.

CHAPTER 7

PALEOANTHROPOLOGY IN HUNGARY

L. KORDOS

Three palaeoanthropological localities are known in Hungary, and will be discussed in this presentation. The oldest is Rudabánya in northeast Hungary. It contains the late Miocene hominoid *Rudapithecus hungaricus*, that can be found with *Anapithecus hernyaki* belonging to the gibbon line. The second is Vértesszölös representing Middle Pleistocene sediments, tooth remains and skull fragments of early Man accompanied by stone tools, archeological layers and evidence of fire. The most important locality of the Neanderthal Man in Hungary is Suba-lyuk in the Bükk Mts, with paleolithic finds and charcoal remains.

THE RUDABANYA HOMINOIDS

At the Rudabánya locality 116 primate finds were gathered between 1967 and 1991, the majority of which are of extreme importance in primate and hominoid evolution. The first item, a mandible fragment, was passed over to Professor M. Kretzoi in 1967 who immediately named it *Rudapithecus hungaricus* (Kretzoi, 1967; 1969). In 1969 the geologist G. Hernyák collected another *Rudapithecus* mandible from Site I (Rud-2). Regular investigations were started in 1970, headed by Prof. M. Kretzoi, sponsored by the Hungarian Central Office of Geology, and carried out, in co-operation, by the Hungarian Geological Institute and the Department for Zoology and Anthropology of the University of Debrecen. In the course of these excavations led by M. Kretzoi, altogether 43 primate finds were found between 1970 and 1975.

Kretzoi distinguished three taxa: *Rudapithecus hungaricus* (Kretzoi, 1969), *Bodvapithecus altipalatus* (Kretzoi, 1974), and *Pliopithecus (Anapithecus) hernyaki* (Kretzoi, 1974).   During this period of excavations the number of primate localities increased to three in Rudabánya, Sites I, II and III.   Prelim-

FIGURE 1
PALAEOANTHROPOLOGICAL LOCALITIES IN HUNGARY

inary reports and evaluations of the rich paleobotanical, ostracod, mollusc, vertebrate and primate material were soon published (Kretzoi, 1974; 1975; 1976a; 1976b; Kretzoi et al. 1976). The collecting work headed by Kretzoi was continued by the Hungarian Geological Institute from 1976-1978, following intensive mining operations and the development of Site I to a nature conservation area. By 1978 a total of 75 primate items were found of which the compressed maxilla and face fragment RUD-71 was the most important (Site II). Kretzoi named this *Rangwapithecus* (*Ataxopithecus*) *serus* (Kretzoi, 1984). Between 1979 and 1984, although no new primate finds were gathered, the rescue excavations, laboratory analyses and geological loggings were continued, headed by Kordos.

In September, 1985, when clearing Site II, from the clay between the two lignite beds, G. Hernyák found *Rudapithecus* skull pieces and teeth/RUD-77 under a coalified tree trunk. (Kordos, 1987; 1988). In 1988 the *Anapithecus hernyaki* skull fragment, (RUD-83), and then in 1989 several teeth, were found at Locality I, Rudabánya (Kordos, 1990).

The village Rudabánya is in Borsod-Abauj-Zemplón county northwest of the town Miskolo. The 7 km long and 1-2 km wide hilly range at the settlement is a Triassic carbonate rock sequence, compressed tectonically from the NW and SE directions. The present day extremely complicated geological structure developed due to a lateral compression that occured during the Middle Miocene. The formation of the Rudabánya hill range was completed by this event. Some 11-12 million years ago, at the boundary of the Sarmatian and Lower Pannonian, a large-scale subsidence started in the Carpathian Basin. As a result, it got filled in by the Pannonian Sea, and then, after the Porta Ferrae area was cut from the sea, by the Pannonian Lake. During the Lower Pannonian, three morphologically different facies could be distinguished in the Borsod Basin: (1) terrestrial, (2) fluviatile-swampy, and (3) lacustrine (Kordos, 1982; 1985). Within the closed terrestrial arch of the Borsod Basin, at a height of 300-320 m., denuded subtropical peneplains were already formed before the Pannonian. The peneplains were deformed due to karstic processes and fluviatile erosion. During the Lower Pannonian (Bodva Stage, i.e., the period of hominoids) the Borsod Basin was a bay of the Lower Pannonian Lake, and had alternating terrestrial karstic areas and swampy-lacustrine subbasins. By the period of the Lower Pannonian, a valley system emerged in the elevated section of the swampy environment that, on the northwest margin of the range was reached by the Pannonian Lake. (Fig. 2).

In the whole valley system the geological formations are similar: at the bottom the ironiferous basement is covered by 2-8 m of thick reddish varigated clay that is lacking in fossil remains, then follows the alternating clay and lignite layers (altogether in a thickness of 4-12 m). The sedimentary sequence is covered by 10-13 m of thick sand that is free from lignite and fossils and these sediments completely fill up the valleys. In the fossiliferous middle part that is rich also in primates, 2-8 lignite beds were formed with a varying thickness of 0.2-1.2 m, and the thickness of the intercalated clay sediments is 1.0-2.5 m.

FIGURE 2
THE PRE-PANNONIAN RELIEF OF THE PENINSULA OF RUBABANYA

1=the extent of Pannonian lacustrine and marshy deposits
2=localities
3=tectonic lines on the margin of the mountains
Kordos, 1990 b

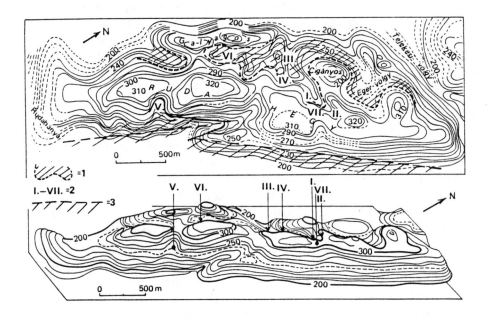

From the geology of the Rudabánya locality two problems are of stratigraphical importance:

1. the fossil localities were deposited in swampy areas in the valleys of a peninsula emerging from the Pannonian Lake. According to our present knowledge no direct lithological connection exists between the rock layers of the valley and basin facies and
2. the lignitic sequence of the Rudabánya valley deposits was formed within a short period of time, and within one biozone.

The Vertebrate biostratigraphical age of *Rudapithecus* can be excellently determined by the rich accessory fauna of the locality. The basic features of the Vertebrate fauna can be best determined based on the *Hipparion* fauna of archaic character (Kretzoi, 1975). Concerning the European Neogene Mammal Zonation, it has not been discussed yet if the Rudabánya fauna can be classified into the MN 9 Zone (Mein, 1971; Kordos, 1985; 1987; Rabeder, 1985; Kretzoi and Pécsi, 1982). In the Middle Paratethys Neogene strigraphy, the Carpathian Basin MN 9 Zone can be correlated with the beginning of the Lower Pannonian while in the Tethys environment, with the Early Vallesian. No radiometric dating has been carried out on the Rudabánya localities because the finds proved to be unsuitable for measurement with the methods intended for application. In the Borsod Basin where the Rudabánya peninsula intruded during the Lower Pannonian, continuous sediment deposition with tuff horizons could be found between the Sarmatian and the Pannonian. Based on several, recently determined radiometric measurements (K-Ar method) the duration of the volcanism of the Tokaj mountains east of Rudabánya was found to reach also into the Lower Pannonian between 14.10-8.94 myr (Balogh, 1984; Balogh and Jámbor, 1987). This means that, based on radiometric data, obtained from the northeastern Hungarian rhyolithic volcanites, the Sarmatian-Pannonian boundary can be placed at 11.5±0.5 myr (Balogh, 1984).

The *Hipparion* data correlate with the radiometric dating. In Kretzoi's (1981) opinion, the rich *Hipparion* occurrence at Rudabánya indicates the first *Hipparion* invasion in the Carpathian Basin. This statement is based on the evolutionary characteristics and the age of the accessory fauna. No earlier occurrence of *Hipparion* has been shown to exist in the Carpathian Basin. A new revision of the *Hipparion* datum (Bernor et al., 1987) has been given as 11.5 myr. This, independent of the local Sarmatian-Pannonian determination, has led to similar results.

Thus, the conclusion can be drawn that the age of the *Rudaithecus* locality is not older than 11.5 myr and probably, it is not younger than 10 myr.

Concerning the systematic determination of the Rudabánya primate fossils, of course, several different opinions have been published. They agree in one point: in the Rudabánya material, two large taxa can undoubtedly be distinguished: (1) *Rudapithecus-Bodvapithecus* group, (2) *Pliopithecus-Rangwapithecus* (Figs. 3-5). Based on the RUD-1 find, denomination of *Rudapithecus hungaricus* was first mentioned by Kretzoi (1967) in a report given to a Hungarian daily paper. Still in the same year, in the course of an international symposium Kretzoi briefly characterized *Rudapithecus hungaricus* (Kretzoi, 1975): "A hominoid of slight build with a short face; high symphyseal, parasagittal, section of premaxillares; wide incisive canal; flat palate; subparabolic dental arch; relatively small I^1 premolars subequal in size; low, brachyodont postcanines without cingula."

These features are still valid for *Rudapithecus hungaricus*.

FIGURE 3
RECONSTRUCTION OF THE *RUDAPITHECUS HUNGARICUS* SKULL
IN LATERAL VIEW, AS ASSEMBLED FROM RUD-77, RUD-12, RUD-17 AND RUD-2 FINDS.

KORDOS, 1987

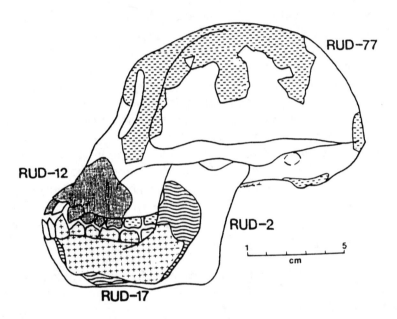

FIGURE 4
A RECONSTRUCTION OF SKULL OF MALE (LEFT) AND FEMALE (RIGHT)
RUDAPITHECUS HUNGARICUS
KORDOS, 1990B

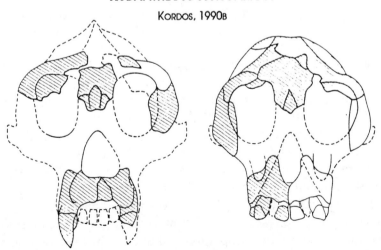

FIGURE 5
A RECONSTRUCTION OF SKULL OF *ANAPITHECUS HERNYAKI* IN FRONTAL AND LATERAL VIEWS

KORDOS, 1990B

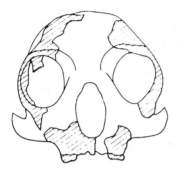 

So far, based on previous studies of the present author, the morphological and phylogenetical relations can be summarized as follows. The linea frontalis form, the presence of the frontal sinus, the vertically enlongated orbits and the lateral view of the face indicate similarity between the skull (KNM-WK 16999) of the male *Afropithecus urkanensis* (Leakey et al., 1988) and *Rudapithecus hungaricus* (RUD-44). A difference is apparent, however, in the pattern and ratio of the teeth, in the maxillary arch, in the smaller orbit size of *Afropithecus*, and in its rather wide interorbital distance and in the mandible structures of the rather robust *Afropithecus* and more gracile *Rudapithecus*. By the reduction of the face of the stratigraphically older *Afropithecus*, the morphology of the RUD-44 can be created, while from the extremely primitive tooth morphology of *Afropithecus* the tooth morphotype of *Rudapithecus* cannot be deduced. The skull of *Turkanopithecus kalakolensis* (Leakey et al., 1988b) (KNM-K 16950) shows similarity with *Rudapithecus* (RUD-77) in the lateral arch of the neocranium, in the reduced face, in the linea temporalis, in the wide interorbital distance, in the D-shaped orbit and in the height of the palate. They are basically different, however, concerning the maxillary arch, the tooth-pattern, the mandible structure and in its size. Both *Turkanopithecus* and *Rudapithecus* are taxa leading to the reduced face where the older (by 7 myrs) *Turkanopithecus'* primitive mandible, maxilla, and tooth pattern show a different morphology from that of *Rudapithecus*.

Proconsul africanus (Walker et al., 1983) skull KNM-RU 7290 is of similar morphology with *Rudapithecus* RUD-77 in respect to the linea frontalis, the presence of a sinus frontale, in lack of the supraorbital tori, in the straight zygomatic arch, the arch of the upper tooth-row, the reduced face, and in the structure of the gracile mandible. The differences are as follows: *Pronconsul africanus* had a shortened while *Rudapithecus* had an elongated neurocranium, and the interorbital distances differ as well as the form of the orbits, the level of the fosa canina, and the tooth patterns. *Proconsul africanus* shows an analogous way of evolution with that of *Rudapithecus*, though some changes exist. The same level of the reduction of the face took place some 6-8 myrs earlier with *P. africanus*.

From the *Proconsul nyanzae* finds, the morphotype of *Rudapithecus* can be deduced if it is assumed that the canines get smaller, the P_3 is rotated, the row sequence of the M_1-M_3 teeth is lessened, and on the teeth M_2-M_3 the cingulum disappears. *Proconsul major*, due to its obvious metric and morphological features, shows no verifiable lineage with *Rudapithecus hungaricus*.

From *Proconsul* (=*Rangwapithecus*) *gordoni* the morphology of *Rudapithecus* can be deduced if the size of the canines decreases, the tooth row slightly increases, the cingulum disappears, and a narrower maxilla develops. In the case of *Proconsul* (=*R.*) *gordoni*, the antero-posterior reduction of the maxilla achieved the level found with *Rudapithecus*.

The maxilla of *Kenyapithecus wickeri* (KNM-FT46) is similar to *Rudapithecus* (RUD-12) except that it is reduced in size and the cingulum is missing. A significant diffence is that with *Kenyapithecus* the fossa canina is rather deep and the teeth are antero-posteriorly strongly pressed. *Kenyapithecus* has a much more reduced facial structure, but based on tooth morphology is a quite similar taxon to *Rudapithecus*. Accordingly, the two genera may not belong to a direct lineage of descendents but a similar ancestor may be supposed.

The identity of the genera *Rudapithecus* and *Sivapithecus* is excluded by several anatomical features. Based on RUD-77 and GSP 15000 samples, these traits are

RUD-77	GSP 15000
lack of supraorbital tori	weakly developed supraorbital tori
elongated skull	shortened skull (reconstruction)
wide interorbital region	narrow interorbital region
shortened face	elongated face
low palate	elevated palate

These differences are shown by the subnasal alveolar morphologies of the African-type RUD-12, and the differing thickness of the tooth-enamel of the Asian-type *Sivapithecus*. Morphological similarites are indicated only by the D-shaped orbits and U-shaped arch of the upper tooth-row (Pilbeam, 1982). In order to see the systematic and phylogenetic aspects, it is important to analyse the relations between *Rudapithecus* of Rudabánya (RUD-77; 10 myr old) and *Lufengpithecus lufengensis* of Lufeng (China, 7 myr old), denoted as P.A. 677. Both fossil skulls belonged to female apes (Kordos, 1987; Wu Rukang, 1986; Lu and Zhao, 1988). The present author was able to study the material in Beijing in 1987 (Kordos, 1988). Summarizing these features, the conclusion can be drawn that there is a great similarity of the skulls in question. There is however, a basic morphological difference between the facial structures of the RUD-77 and P.A. 677 finds, and thus between the taxa *Rudapithecus hungaricus* and *Lufengpithecus lufengensis*. The older early apes of Rudabánya (European origin) and the younger Lufeng species (SE Asian origin) unanimously bear the features of an evolutionary epoch when the development of the cranium had slowed down, and the facial structure went through important morphological and phylogenetic changes. On the above basis a systematic differentiation of the two types at generic level is justified.

At the present time in our study, we can make statements only about the skull of *Ouranopithecus* (de Bonis et al., 1990). We have observed that, in addition to great differences in size, the find from Macedonia (XIR-1) significantly differs from *Rudapithecus* in the structure of the region around the orbit, in the orbit, in the interorbital distance, and in the construction of the maxillary arch. Concerning the most recently (1991) discovered *Dryopithecus* skull (Can Lllobateres, Spain), no sufficient data are available. A comparison of *Rudapithecus* with other hominoid remains indicates that in the transformation of the skulls, different anatomical processes took place with a different evolutionary speed, in a mosaic type manner, and heterogeneously. Changing of the tooth pattern from one to the other at a morphotype scale took some millions of years. That is why the tooth morphotype analyses seem to be applicable for determining the chronological hiatus between the individual finds.

After studying the following features of the African and Eurasian hominoids and hominids — the morphological features of M_3/cingulum, fovea anterior, extraconulus between the metaconid and entoconid, extra conulus between the entocoid and hypoconulid, fovea posterior — the evolutionary line of several tooth morphotypes could be modelled. Based on all of this, the most probable hypothesis seems to be that the African ancestor of *Rudapithecus* should be sought in the *Proconsul africanus-nyanzae* group. At the same time it can be assumed that the European *Dryopithecus* (including *Rudapithecus*) can be interpreted as a dentally morphological sister group of the *Sivapithecus* (6. ábra).

The anatomical structures of *Rudapithecus* support the hypothesis that both the origin of Man and the chimpanzee and gorilla can be derived from it.

In spite of this fact, however, the zoogeological occurrences do not testify to these phylogenetic relations. In addition to *Rudapithecus hungaricus*, there is another fossil ape called *Anapithecus hernyaki*. A skull fragment of this extinct giant gibbon was found in 1986, and also several teeth are known. Among the Pliopithecinae, *Epipliopithecus vindobonensis* (Zapfe, 1960) morphologically and allometrically differs from *Anapithecus*. *Laccopithecus robustus* from China (Pan, 1988) is morphologically more similar to *Anapithecus* than *Epipliopithecus*. Based on tooth morphology, among the European *Pliopithecus*, the closest phylogenetic relation (Fig. 7) may be supposed with *P. antiquus* of Göriach (Hürzeler, 1954). African relations of the European *Pliopithecus* may be most probably sought among *Dendropithecus*.

FIGURE 6
EVALUATION OF THE PHYLOGENETIC RELATIONS OBTAINED BY M3 TOOTH MORPHOTYPE ANALYSIS WITH REGARD TO OTHER FEATURES OF THE HOMINOIDS

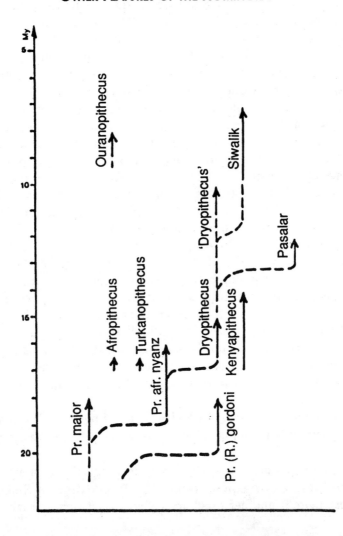

FIGURE 7
PHYLOGENETICAL RELATIONS OF SOME LESSER APES
KORDOS, 1989A

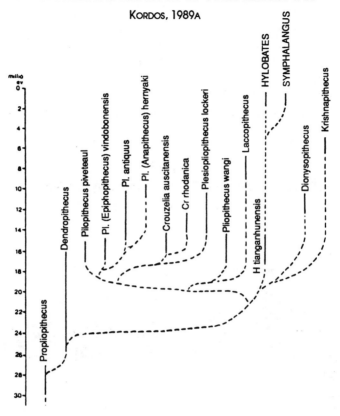

THE VÉRTESSZÖLÖS EARLY MAN

The locality is at the village of the same name at the southern foot of the Gerecse Mountains (NW Hungary). From the freshwater limestone sequences explored by mining, different sporadic vertebrate fossil remains have been recovered from the beginning of our century. The Early Man Site was explored after a students' tour aimed at collecting pebbles and bones in 1962. The excavations led by L. Vértes came to a stop in 1968 because of his death. At present the locality is an open-air exhibition under the supervision of the Hungarian National Museum. The results achieved so far were published in a monograph edited by M. Kretzoi and V.T. Dobosi (1990).

The travertine complex of Vértesszölös was deposited upon the terrace of the Átalér brook at a relative height of 60 m with a thickness of 8-10 m. Based on geomorphological and geological data the travertine enclosing the locality was formed during the Günz-Mindel Interglacial, at the beginning of the Mindel Glacial (Fig. 8). The cultural layers were deposited in a travertine basin in 3 levels, that, with loess intersections, were covered by another travertine deposit. In the quarry region several sites were differentiated as follows: Site I (*Homo erectus* settlement), Site Ia (Talpa), Site II (Carnivora, or

Jánossy's microfauna locality), Site III (Foot print) and Site VI (Molluscs). Of these sites, III is the oldest with footprints of *Ursus daningeri*, *Stephanorhinus etruscus*, Cervidae and *Bison*. Among the small-vertebrate fauna a great number of cold-bearing species occur like *Microtus gregalis*, *Microtus conjungens* and *Ochotona sp*. Sites I, Ia and VI are of similar, warmer fauna of the forest type. Among the large mammals, *Equus, Bison*, and *Stephanorhinus etruscus* occur frequently. The species characteristic of the small mammals are the *Trogontherium schmerlingi, Pliomys episcopalis, Apodemus sylvaticus* and *Budamys synanthropus*. Site II represents fauna of cold climates and is rich in birds, bats and small and large mammals.

FIGURE 8
A GEOLOGICAL PROFILE OF THE VÉRTESSZÖLÖS EARLY MAN SITE
AFTER HENNIG ET AL., 1983

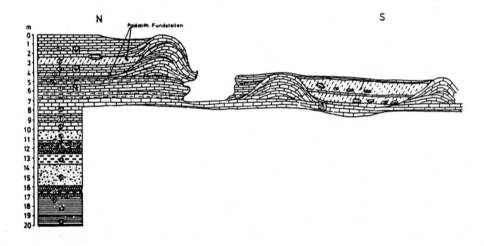

Based on the fossil vertebrates, Site III was put into the Lower Biharium's Tarkö phase, while the Sites I, Ia and VI were designated Upper Biharium Vértesszölös phase (Kretzoi, 1990). According to Kretzoi and Jánossy (1990), Site II belongs to the Uppony phase that followed the Vértesszölös phase. Revising the Vertebrate faunas of the Vértesszölös locality (Kordos, 1991 in press) it can be concluded that no biostratigraphically significant differences can be found among the sites. The systematic revision of the Arvicolida fauna is, however, essential because formerly it was interpreted by two experts in a completely different manner. At the same time it seems necessary to adopt the Kretzoi-Jánossy biochronological system to the Middle European biostratigraphic convention (Fejfar and Heinrich, 1990). On this basis all the species of the Vértesszölös fauna are part of the *Arvicola cantiana* Zone. They can be differentiated by paleoclimatological data, as there was an early cool epoch (Vsz. II, III) and following this, a warmer, forest period (Vsz. I, Ia, VI). The travertine formation, with loess sediments is closed by another cool period without Vertebrate remains. Biostratigraphically the Vértesszölös faunas are bordered by the *Lagurus transiens-Arvicola contiana* Zone (Vsz. Sites II and III; Elsteriana cca. 350 kyr) from the bottom. The majority (Vsz. Sites Ia, I, VI) belong to the *Arvicola cantiana* Partial Range Zone / Holsteinian and the warm period of the Early Saalian; 320-250 kyr. These faunas, however, do not include the typical species of the next zone (*Lagurus lagurus* Partial Zone). The radiometric dating of the different samples performed by several experts (Cherdyntsev et al., 1965; Osmond, 1990; Hennig et al., 1983) do not contradict this correlation. The corrected data supplied by Schwarcz and Latham (1984) indicate, however, a much younger age (225-198 kyr).

FIGURE 9
THE VÉRTESSZÖLÖS II OCCIPITAL BONE
THOMA, 1966

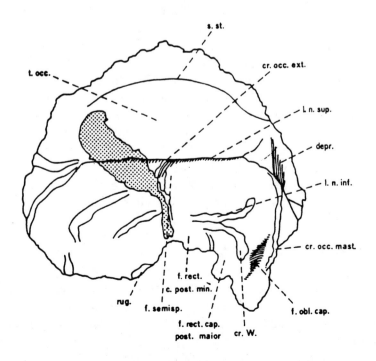

According to Thoma (1966; 1967; 1990), the Early Man remains of Vértesszölös represent two assemblages, i.e., the Vértesszölös I and II finds. Vértesszölös I, collected from the first cultural layer in 1965, consists of 4 tooth fragments belonging to one individual (ca. 7 years old). Based on their primitive morphology they may be related to *Sinanthropus*. Vértesszölös II, a human occipital bone (Fig. 9) from the topographical horizon corresponding to Layer I, was excavated at an 8 m distance from freshwater limestone in 1965. According to Thoma (Thoma, 1990):

> The find consists of the squama occipitalis which was diagonally broken into two parts by the blasting; the freshly broken edges articulate well. Its lambda region is postmortally deformed in a vertical direction, but this deformation can be precisely corrected. The area surrounding the foramen magnum was forcefully broken: the position of the opisthion could be reconstructed since a portion of the thickened boned / limbus occipitalis / surrounding the foramen magnum had survived. The bone is robust, very broad and the intact parts of the lambdoid suture are open. This implies that it belonged to a young, adult male.

The cranial capacity, was determined by Thoma as 1300 cm. According to his interpretation, based on anatomical and metric features, the Vértesszölös Man is modern, while in general it is a robust *Archanthropus*. Because of the transitional characteristics, Thoma gave the name *Homo erectus seu sapiens palaeohungaricus n.ssp.* to differentiate the Vértesszölös II find from other early types. Thoma's conclusions were discussed by Wolpoff (1971), first of all because of the calculation of the cranial capacity. According to his conclusions the Vértesszölös II find belongs to the *Homo erectus* group and is not similar to the Swanscombe find. Thoma, in his reply (1978), remarked that Wolpoff's lambda location is erroneous, and insisted that he did give the correct value for the cranial capacity. Since finding Vértessszölös II, several new Early Man items have been found in Europe, but due to the decade long delay of the publication of the monograph, no revision could be made. Hopefully, in the near future since the monograph was published in 1990, a revised study of this important work can be carried out.

THE NEANDERTHAL MAN OF SUBA-LYUK

Suba-lyuk is a cave locality in northeast Hungary at the southern foot of the Bükk Mts. Archeological excavations started in 1932, and in that year bone remains of two Neanderthal men were found from the 11th layer at the entrance section. Results of the excavations were published in a monograph (Kadic, 1940).

FIGURE 10

DISTRIBUTION OF THE NEANDERTHAL BONES IN THE SEDIMENT OF SUBA-LYUK CAVE WITH THE SITE INDICATION OF THE RECENTLY FOUND M₂

AFTER KADIC, 1940

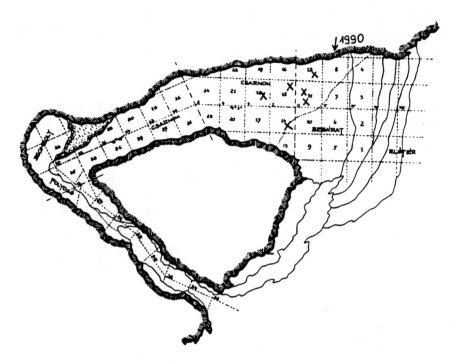

Altogether 18 layers have been distinguished in the cave (Fig. 10). According to the original interpretation the three lowermost layers belong to the Riss-Würmian interglacial, and the next three layers are transitional. Layers 7-16 were deposited next as "upper layers" and were determined to be Würmian I. Beside the Mousterian type stone tools in the lower layers (1-3) are found *Capra ibex*, while in the upper ones (7-16), the cave-bear was the dominant fauna, accompanied by steppe-type small mammal faunas. From the level corresponding to the 11th layer D. Jánossy later (1960) indicated from samples of the cracks in the wall, 14 small mammal species with *Lagurus lagurus* steppe species as dominant. At that time it could not be determined with what upper layer this fauna could be matched. That is why Vértes (1965) bitterly remarked that "We are to give up hope to prepare a more exact stratification on the basis of the faunistic results."

Based on archeological and sedimentological data Vértes (1965) put the age of the upper cultural layer (it contained also the Neanderthal finds), to somewhere in the Würmian I maximum. To improve this unstable chronological situation the elaboration of a new biostratigraphical system for the zoo-geology of the Arvicolidae of Central Europe was carried out

(Kordos, 1990; 1991 in press). Also finding a new fossil assemblage was most helpful. In 1990 A. Ringer, archeologist, and P. Solt, preparator, took samples from the same locality. In the small sample, together with the dominant *Lagurus lagurus* small mammals, the left M_2 tooth fragment of a Neanderthal was also found (Fig. 11). As a result, the exact stratigraphic determination became possible since it could not be proved that Jánossy's sample and its vertebrate fauna are identical with the layer bearing the Neanderthal find (Kordos, 1990; Kordos, Krolopp and Ringer, 1991).

FIGURE 11
FRAGMENT OF THE LEFT SECOND MOLAR OF THE NEANDERTHAL MAN, FOUND IN THE SUBA-LYUK CAVE, 1990

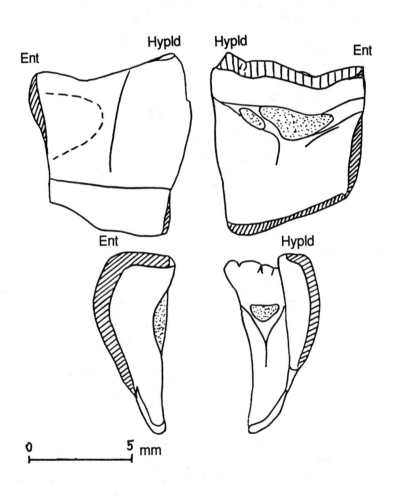

Among the Arvicolidae occurring in the 11th layer the dominant *Lagurus lagurus* migrated to Central Europe from the East, and is found there by the end of the Middle Pleistocene. Its sporadic occurrence can be traced in the Carpathian Basin and in the territory of the present Poland up to the last glacial period of the Wurmian. During this period its dominant presence can be correlated at about 60 kyr when they invaded Europe from two routes. One of them was from the southern part of the Russian Plain arriving at the Carpathian Basin following the Danube line; and they stopped there. The other route led to the North, and across the German-Polish Plain until they got to Belgium and France. The presence of this species, together with other mammals from the steppe environment, indicates one of the most characteristic, biostratigraphical levels of the Late Pleistocene. Biostratigraphically, the layer is covered by *Microtys gregalis - Lagurus lagurus* Zone. The 11th layer of the Suba-lyuk, together with the Neanderthal remains, was deposited during this event. The anthropological finds of Suba-lyuk were found scattered about in an area of some 20 m. This seemingly contradicts the hypothesis that the remains were buried. At the same time, it is also possible that the bones of the originally buried corpses were removed by posterior soil slides, or other natural climatic movements.

One of the Neandertthal finds comes from an elderly woman but it is possible that the remains belong to different individuals. Its anatomical description and the contemporary level evaluation together with the other find, a child's skull, is given by Bartucz (1940) and Szabó (1935). The dentition of the Suba-lyuk child was later studied by Thoma (1963) who interpreted it to be a classic Neanderthal individual that reflects several archaic features.

By summarizing briefly the three important fossil assemblages reflecting three extremely different stages of human evolution, it can be concluded that the main thrust of research in 1991 is on the excavations at Rudabánya and on the complex, international evaluation of the finds. Also in this year, and after a long delay, the Vértesszölös monograph was finally published (Kordos, 1991). In the case of Suba-lyuk no important anthropological investigations have been carried out at present, but we could determine the biostratigraphical position of the upper layers, and the archeologists are intensively studying the Paleolithic finds.

REFERENCES

Balogh, K. 1984. Methods and results of K/AR method in Hungary. Ph.D. Thesis, Debrecan (in Hungarian).
_____. and A. Jámbor. 1987. Radiometriche Daten zur Charakteristik postsarmatischer Ablagerungen in Ungarn. *Földt. Int. Évk.* LXIX., 27-36.

Bartucz, L. 1940. A Mussolini barlang ösembere. *Geol. Hun. Ser. Pal.*, 14.49-99.

Bernor, R.L., M. Brunet, L. Ginsburg, P. Mein, M. Pickford, F. Rögl, S. Sen, F. Steininger and H. Thomas. 1987. A consideration of some major topics concerning 0_1d World Miocene Mammalian chronology, migrations and paleogeography. *Geobisos*, 20:4, 431-439.

Bonis de, L. G. Bouvrain, D. Geraads, and G. Koufos. 1990. New hominid skull material from the late Miocene of Macedonia in Northern Greece. *Nature*, 345: 712-714.

Cherdyntsev, V.V., I. Kazachevskii and E.A. Kuzmina. 1965. Dating of Pleistocene carbonate formations by the uranium and thorium isotopes. *Geochemistry International*, 2, 794-801.

Fejfar, O., W.D. Heinrich. 1990. Muroid Rodent Biochronology of the Neogene and Quaternary in Europe. Lindsay, E.M. et al. (ed.). *European Neogene Mammal Chronology*, 91-117, Plenum Press, New York.

Hennig, G.J., R. Grün, K. Brunnacker and M. Pécsi. 1983. Th-230/U-234 sowie ESR-Alerbestimmungen einiger Travertine in Ungarn. *Eiszeitalter u. Gegenwart*, 33: 9-19.

Hürzeler, I. 1954. Contribution a l'odontologal et la phylogenese du genre Pliopithecus Gervais. *Annales de Paléont.*, 40, 1-63.

Kadic, O. ed. 1940. A cserépfalui Mussolini barland. Subalyuk. *Geol. Hung. Ser. Pal.* 14.

Kordos, L. 1982. The Prehominid Locality of Rudabánya (NE Hungary) and its Neighbourhood: A Palaeogeographis Reconstruction. *Földt. Int. Évi Jel.* 1980-ról., 395-406.

_____. 1985a. Environmental reconstruction for Prehominids of Rudabánya, NE Hungary. *Schr. zur Ur und Frühgeschichte*, 82-85, Berlin.

_____. 1985b. *Neogene Vertegrate Biostratigraphy in Hungary*. Proc. VIII. RCMS Congr., Földtani Int. Évkönyve, LXX., 393-396.

————. 1987. Description and reconstruction of the skull of *Rudapithecus hungaricus hungaricus*, Kretzoi. Mammalia. *Annls. Hst. Nat. Mus. Natan. Hung.*, 77-88.

————. 1988a. Comparison of Early primate skulls from Rudabánya. Hungary and Lufeng, China. *Anthropologica Hungarica*, 20: 9-22.

————. 1988b. *Rudapithecus* skull finds from the Lower Pannonian of Rudabánya. *Földt. Int. Évi Jel.*, 1986-ól, 127-154.

————. 1989. Az utolsó európai gibbon. *Természet Világa*, 120:11, 525-527.

————. 1989. Az emberré válás rudabányai ösmajmai. *Scientific American*, Hungarian ed., 5:8, 61-63.

————. 1990a. Ujabb rudabányai leltek és as emberré válás sokasodó elméletei. *Magyar Tudomány*, 1, 9-14.

————. 1990b. Hominoidea locality, Rudabánya, Rudabánya Mountains. *Magyarország geológiai alapszelvényei*, MÁFI ed., 1-6.

L. 1990c. Biostratigraphy and Paleoecology of the Middle European Late Quaternary. Third Symp. Upper Palaeolithic, Mesolithic and Neolithic Population of Europe and the Mediterarean Basin, Abstract, Budapest.

————. 1990d. The evolution of Upper Pleistocene voles in Central Europe. Int. Symp. Evol. Phyl. Biostr. Arvicolids, Pfeil-Verlag, 275-284.

————. 1991. Revised Biostratigraphy of the Early Man Site at Vértessölös, Hungary. 4th Internat. Senckenberg Conf., *"100 Years of Pithecanthropus - The Homo erectus Problem."* Frankfurt am Main, in press.

————., E. Krolopp and A. Ringer. 1991. Le complexe Bábonyien-Szélétien dans son cadre chronologique-paléoecologique. *Coll. Commen. Internat.* Miskolc 1891-1991, Abstract.

Kretzoi, M. 1967. Magyar Nemzet. Oct. 1. report.

————. 1969. Geschichte der Primaten und der hominisation. *Symp. Biol. Hung.* 9., 3-11.

————. 1974. Towards Hominization. *Anthrop. Közlem.* 18, 121-128.

————. 1975. New ramapithecines and Pliopithecus from the Lower Pliocene of Rudabánya in north-eastern Hungary. *Nature*, 257, 578-581.

————. 1976a. Die hominisation und die Australopithecinen. *Anthrop. Közlem.* 20, 3-11.

————. 1976b. Die Ramapithecinen von Rudabánya in Nordost-Ungarn. UISPP IXe Congr. Coll. VI. 68.

————. 1981. Wirbeltier-indexformen im Ungarischen Jungneozoikum, Hipparion. *Földt. Int. Évi Jel.* 1981-rol. 513-521.

————. 1984. New hominoid from Rudabánya. *Anthrop. Közlem.* 28., 91-96.

————. 1990. Vertebrate Microfauna of the Vértesszölös Travertine. In Kretzoi, M. and V.T. Dobosi (eds.). *Vértesszölös Site, Man and Culture,*

241-247, Akadémiai Kiadó, Budapest.

———. and V.T. Dobosi eds. 1990. *Vértesszoölös Site, Man and Culture.* Akadémiai Kiadó, Budapest.

———., E. Krolopp, H. Lorincz and I. Pálfalvy. 1976. Flora, Fauna und Stratigraphische Lage der Unterpannonischen Prahominiden-Funstelle von Rudabánya. NO-Ungarn. *Földt. Int. Évi Jel.* 1974-ról. 365-394.

———. and M. Pécsi. 1982. Pliocene and Pleistocene Development and Chronology of the Pannonian Basin. *Földr. Közlem.* 30. 106. 4, 300-326.

Leakey, R.E., M.G. Leakey and A.C. Walker. 1988a. Morphology of *Turkanopithecus kalakolensis* from Kenya. *Am. J. Phys. Anthrop.,* 76: 277-288.

———., M.G. Leakey and A.C. Walker. 1988b. Morphology of *Afropithecus turkanensis* from Kenya. *Am. J. Phys. Anthrop.,* 76: 289-307.

Lu Qingwu and Zhao Zhongyi. 1988. The Reconstruction of the Head of the Female Lufeng Ape. *Acta Anthrop. Sinica,* 7:1, 9-16.

Mein, P. 1971. Biozonation du Neogene Mediterranéenne. Table.

Osmond, J.K. 1990. Th$^{230/234}$ Dating of Vértesszolos Site, Man and Culture. 545, *Akadémiai kiadó,* Budapest.

Pan Yuerong. 1988. Small fossil primates from Lufeng, a latest Miocene site in Yunnan Province, China. *J. Human Evolution,* 17, 359-366.

Pilbeam, D. 1982. New hominoid skull material from the Miocene of Pakistan. *Nature* 295: 232-234.

Rabeder, G. 1985. Die Saugetiere des Pannonien. Chronostratigraphie und Neostratotypen. M_6 Pannonien. Eds. Papp, A., A. Jámbor and F.F. Steininger. 440-463. *Akadémiai Kiadó,* Budapest.

Schwarcz, H.P. and A.G. Latham. 1984. Uranium-series Age Determination of Travertines from the Site of Vértesszöllös, Hungary. *J. Archaeol. Sciences,* 11, 327-336.

Szabó, J. 1935. L'homme moustérian de la grotte Mussolini Hongrie. Étude de la mandibule. *Bull. Mém. Soc. d'Anthrop.* Paris. 6, sér. VIII.

Thoma, A. 1963. The dentition of the Subalyuk Neanderthal child. *Z. Morph. Anthrop.,* 54:2, 127-150.

———. 1966. L'Occipital de l'Homme minélien de Vértesszöllös. *L'Anthropologie,* 70, 495-534.

———. 1967. Human Teeth from the Lower Palaeolithic of Hungary. *Z. Morph. Anthrop.,* 58, 152-180.

———. Some Notes on Wolpoff's Notes on the Vértesszölös Occipital. *J. Human Evolution,* 7, 323-325.

———. 1990. Human Tooth and Bone Remains from Vértesszölös Site. Man and Culture, 253-262. *Akadémiai Kiadó,* Budapest.

Walker, A., D. Falk, R. Smith and M. Pickford. 1983. The Skull of *Proconsul Africanus* Reconstruction and Cranial Capacity. *Nature,* Vol. 305, No.

5934, 525-527.

Wolpoff, M.H. 1971. Is Vértesszöllös II an Occipital of European *Homo Erectus? Nature*, 5312, 567-568.

Zapfe, H. 1960. *Die Primatenfunds aus der miozanen spaltanfüllung von Neudorf an der March*. Devinská Nová Ves. Tschechoslowakei. Schweizer. Palaeont. Abh. 78, 1-293.

CHAPTER 8

INTERDISCIPLINARY STUDIES IN THE PROJECT "MENSCHWERDUNG"
(ORIGINS AND EVOLUTION OF HUMANS)

H. ULLRICH

In eastern Germany physical anthropology was limited to 3 institutes: the Anthropological Institute of the Humboldt-University in Berlin, the Friedrich-Schiller-University in Jena and the Institute of Archaeology of the Academy of Sciences with its anthropological laboratory in Berlin. Research has mainly been done into the origin and evolution of humans, the Neolithic to Mediaeval populations, developmental processes as well as into human variability and human genetics. Medical anthropology, the application of anthropological results to industrial production, anthropological studies in sports and in nutritional problems of mankind were also fields of research. There is only one full professor of anthropology, always the head of an anthropological institute in eastern Germany. There was neither a possibility to have an ordinary study in anthropology nor to become a full anthropologist. Only students of biology could specialize and have done their graduate or doctoral thesis in the field of anthropology. Lectures in anthropology were given to students of medicine, psychology, archaeology and, occasionally, to students of biology. Four universities in eastern Germany (Rostock, Greifswald, Halle, Leipzig) have no anthropological institute at their disposal and an anthropological journal does not exist. The Anthropological Society of eastern Germany, founded in 1960 within the Biological Society of the GDR, has all the time been very active and has organized more than 20 conferences with participants from eastern and western countries.

Physical anthropology was not promoted by the government of the German Democratic Republic. The government was not interested in it because physical anthropology was considered a science not immediately connected with industrial production and not supporting the ideological background of the government.

INTERDISCIPLINARY WORKING GROUP

Human evolutionary research, limited to the origins and evolution of Pleistocene or Palaeolithic humans, was carried out in eastern Germany only by the Archaeological Museum in Weimar and Halle, the Institute of Quaternary Palaeontology in Weimar and the Archaeological Institute of the Academy of Sciences in Berlin. Since our country has provided only a few important human fossils and sites of Palaeolithic Man, most research done in the last few decades has been focussed on the theoretical aspects of the evolution of Man and human society from multiple and interdisciplinary perspectives. There has been a great interest shown by archaeologists, anthropologists, palaeontologists and others to discuss these problems in greater detail.

Multi- and interdisciplinary discussions on general aspects and problems of the origins and evolution of Man and human society started in eastern Germany in the late 1960s. Archaeologists were engaged in discussions on the periodization of human evolution and ancient society (Feustel, 1968). These discussions were stimulated by the journal *Ethnographisch-Archäologische Zeitschrift* publishing the papers in vol. 14 1973, 55-133 and later. In 1977 an interdisciplinary conference "Die Entstehung des Menschen und der menschlichen Gesellschaft" (Origins of Man and human society) was organized by archaeologists (Fachgruppe Ur-und Frühgeschichte der Historiker-Geselischaft) in Frankfurt/Oder. This conference (proceedings published in 1980 by Schlette) clearly demonstrated that the totality and complexity of the human evolutionary process could only be analysed and recognized by multi- and interdisciplinary approaches. Physical anthropologists as well as archaeologists are unable to comprehend or outline the whole process. This can only be done by co-operation of scholars, working on issues of human evolution and ancient society from different disciplines and from multiple perspectives.

An interdisciplinary working group "Menschwerdung," (anthropogenesis) headed by H. Ullrich, was therefore founded at the Zentralinstitut für Alte Geschichte und Archäologie der Akademie der Wissenschaften in Berlin in 1977. Some 30 specialists, all from eastern Germany, from various disciplines (physical anthropology, archaeology, ethnography, ethology, evolutionary biology, geophysics, history, palaeontology, pedagogics, philosophy, psychology etc.) working on aspects or interested in problems of the *Menschwerdungsprozeß* process of anthropogeny, started to co-operate. From the early beginning there was agreement among all scholars of the working group that humans have to be considered biotic and socio-cultural beings and that the evolution of Man and human society has to be considered a unity. Menschwerdung (anthropogenesis) is therefore a multi-leveled and complex process starting with the origins and evolution of humans and human society as well as their interactions and finishing with the origins of anatomically modern Man (*Homo sapiens sapiens*) and the emergence of racial groups (negrids, europids, mongolids, australids) as well as the establishment of the ancient society of hunter gatherers at the end of the Pleistocene.

The aim of this interdisciplinary working group was not only to discuss facts and theoretical aspects of the origins and evolution of humans, but also to prepare manuscripts for the book "Menschwerdung," an overall view of the hominization process from multiple and interdisciplinary perspectives. There were heated and controversial discussions on this project that culminated in the question, whether or not scientists in eastern Germany should start on such a project, because they were far away from the centre of the origin of Man and the main excavations and fossils in Africa and not directly engaged in those excavations and researches. There were pros and cons and the majority of the working group were hesitant. First of all, we had to prove the prerequisites for our own contributions made by the group members to the hominization process which was based on the local sites in eastern Germany and on the general aspects and problems of human evolution.

LOCAL SITES

At the Middle Palaeolithic open occupation site Weimar-Ehringsdorf in Thuringia, dated about 120,000 years, intensive excavations were started by the Museum für Ur und Frühgeschichte Thüringens in Weimar in the early 1950s. The results were published by Behm-Blancke (1960). He was also the first to describe new human remains (parietals, femur) discovered at that site between 1909-1913 and to present many facts about the cultural activity of the pre-Neandertals. Geological fieldwork and investigations into the palaeontological material from Weimar-Ehringsdorf were done in the 1960s and resulted in an international colloquium "Das Pleistozän von Weimar-Ehringsdorf" organized by the Institut für Quartärpaläontologie in Weimar in 1968. The proceedings were published in two volumes edited by Kahlke (1974; 1975). At the Weimar-Ehringsdorf colloquium some scholars first suggested that the site might be older than Eem, the last interglacial. New palaeontological results and radio-metric dating have strengthened this suggestion and have made an age of 210,000 years BP (warm period in the R1B glacial) possible (Jäger & Heinrich, 1982; Feustel, 1983). The human remains from Weimar-Ehringdorf have been restudied by Vlček (1985), and he has suggested that the pre-Neandertals had an anatomically modern man-like skull.

Geological fieldwork and palaeontological studies were carried out also at the Middle Palaeolithic site Taubach near Weimar by the Institut für Quartärpaläontologie. An international colloquium on that site, dated about

110,000 years ago, was held in 1972 and the results were published by Kahlke (1977). Other important Pleistocene sites in eastern Germany (Voigtstedt, Süßenborn, Burgtonna, Weimar) have also been the subject of intensive geological fieldwork and palaeontological research by an international team (Kahlke, 1978; 1984).

Human remains from the Upper Palaeolithic (Magdalenian) site of Döbritz (Kniegrotte, Urdhöhle) in Thuringia were studied by Grimm & Ullrich (1965), the archaeological material by Feustel (Feustel et al., 1971; Feustel, 1974). Bones with modifications resulting from manipulations on humans corpses were very probably buried (Ullrich, 1975).

Only a few new Palaeolithic sites have been excavated in eastern Germany. The Middle Palaeolithic sites Markkleeberg near Leipzig (Baumann & Mania, 1983) and Königsaue, near Aschersleben (Mania & Toepfer, 1973), and the Upper Palaeolithic localities Oelknitz (Musil, 1985), Gleina and Teufelsbrücke (Feustel, 1980) in Thuringia have to be mentioned. Some years ago the nearly complete skeleton of a butchered *Palaeoloxodon* from the Eem interglacial was discovered and excavated near Gröbern, Kr. Gräfenhainichen. From all these sites no human remains are known.

A new *Homo erectus* site was discovered by D. Mania in 1969 near Bilzingsleben in Thuringia. Extensive excavations by the Landesmuseum für Vorgeschichte Halle/Saale started immediately and made this site most famous. Situated underneath a travertine layer, a Lower Palaeolithic archaeological horizon represents an open occupation site from the Middle Pleistocene (Holstein complex) with an age of about 350,000 - 300,000 years. The open air site was situated on the shore of a small lake in the vicinity of the mouth of a small stream. An area of about 900 m² of the living floor has been excavated by D. Mania. The most important results of the continuing excavations are 11 skull fragments and 7 teeth of 3-5 late representatives of *Homo erectus* in Middle Europe as well as thousands of animal bones and artifacts, that enable reconstructions to be made not only of the environment, but most importantly of the activity of the *Homo erectus* group on the occupation site. Three oval or circular structures evidenced by large bones and stones represent three simple dwellings. In front of each dwelling, or nearby, a fireplace and two workshops, as indicated by centrally placed anvils, were discovered. Further workshops with worked stones, bones, antlers and wood were situated in the large area between and away from the dwelling structures. Another activity zone was located on the shore. In 1986 at a distance of 5-8 m from the dwelling structures an obviously roughly circular floor, free of waste material, half-finished artifacts and tools, became visible. This floor extended more than 9 x 6 m in 1990, but excavations of this floor are continuing. It consists of a pavement of compactly packed small pebbles and bone debris mainly, but also of calcified wood. Near the periphery on this pavement a stone anvil, covered with scars demonstrating that bones had been crushed on it, has been the only evidence of activity so far. Nearby a fragment of a left human parietal bone was found. The former function of this pavement is still unknown, but Mania is tentatively suggesting a place of special cultural activities within the occupation site.

Homo erectus from Bilzingsleben hunted large animals. He made large pebble tools as well as small silex artifacts, large bone tools, percussion implements from antlers and also wooden tools. The surface of four bone tools connected with workshops is covered with bunches of parallel engraved lines — intentional marks (Mania & Mania, 1988). The rhythmic sequence of these linear marks may point together with other facts — there might also be "signs" and an animal engraving on bones (see Behm-Blancke, 1987), but this engraving has recently been doubted by the excavator — to the suggestion that late *Homo erectus* was able to communicate by language and to possess abstract thought and also had a simple concept of the world as well as of life and death (Ullrich, Ms.). The human remains found in Bilzingsleben may have been closely connected with mortuary rites celebrated in or outside the occupation site. Vlček has carefully studied the anthropological aspects of these remains, belonging obviously to 3-5 individuals, and described them as *Homo erectus bilzingslebenensis* (Vlček, 1978; Mania and Vlček, 1987).

The rich material discovered in Bilzingsleben has intensively been studied by an interdisciplinary and international team of archaeologists, anthroplologists, geologists and scholars of other disciplines. They have presented and discussed their results at four Bilzingsleben colloquia entitled "*Homo erectus* - seine Kultur und Umwelt." The fourth was held in 1987. The results have been published in issues of the journal "Ethnographisch-Archäologische Zeitschrift," in three Bilzingsleben monographs (Mania et al., 1980; Mai et al., 1983; Mania & Weber, 1986) and in numerous papers (for a nearly complete bibliography on Bilzingsleben see Ullrich, 1990b). A book published by Mania (1990) summarized the results of this important excavation.

DISCIPLINARY RESEARCH

Disciplinary research on general aspects and problems of the origins and evolution of humans was mainly done by anthropologists, archaeologists, ethnographers, ethologists, evolutionary biologists, philosophers and psychologists in eastern Germany. In this paper only some important results and aspects can be mentioned.

PHYSICAL ANTHROPOLOGY

Since 1977 fossil human remains from about 80 Palaeolithic sites in Europe, representing more than 200 individuals, have been studied with respect to bone modifications, mortuary ceremonies, skull-cult and cannibalism (Ullrich, 1978; 1979 a, b; 1984; 1986; in press; Ms.). The results have shown that bone modifications caused by Man with tools (cutmarks, scraping marks, percussion marks, disarticulation marks, etc.) were very frequently found on fossil human bones from the Lower, Middle and Upper Palaeolithic. These bone modifications resulted from interference in human corpses and bones of the deceased — a frequent and wide-spread practice in Palaeolithic times, obviously connected with ritual or cult ceremonies of the dead.

Mortuary practices might have been carried out in various ways during the Palaeolithic. We do not know what Palaeolithic man usually did with his dead and where they were left, but we have bones of "favoured" dead and these remains, in connection with their archaeological context, can tell us something about mortuary practice for favoured dead in Palaeolithic times. It is very likely that most individuals died away from their temporarily and seasonally occupied home bases and that their corpses were never brought back. Human interference in corpses (defleshing, dismemberment, bone fragmentation) was obviously carried out only on favoured dead and at the place of the death, resulting in disarticulated and broken bones. Disarticulated and broken bones of the deceased, especially of the skull, were of great importance to Palaeolithic Man for further mortuary ceremonies within the whole group at the occupation site. After having completed ceremonies for favoured deceased, the disarticulated and broken bones were simply thrown away, deposited intentionally or buried at special places within the home base.

Most of the human bones known from Palaeolithic sites in Europe were obviously bones of favoured dead and handled in such a way. From the European Palaeolithic human bones, a minimum number of 826 individuals are known from 330 sites: 43 from Lower, 258 from Middle and 525 individuals from Upper Palaeolithic. All Lower Palaeolithic, 94.6% of the Middle Palaeolithic and 85.0% of the Upper Palaeolithic individuals were represented by only a few bones (or teeth), mostly 1-2 bones! It is most unlikely that these bones were derived from complete skeletons/corpses buried at the place where the bones were found — an explanation very often given by archaeologists, but also by other anthropologists. These bones were taken to the occupation site by Palaeolithic Man in connection with mortuary ceremonies and placed, deposited or buried there. It is evident that skull remains were most important within the mortuary rites. Cranial remains are known from 95% of the Palaeolithic individuals, postcranial remains only from 45% of the individuals. Obviously there was a skull-cult in Palaeolithic times.

Burials indicating a stronger relation between life and death were associated with more definite rituals. It is known that there are burials of cleaned bones, of the skull and the head, of the skeleton and of the entire corpse. Almost complete or fully complete skeletons have been excavated from only 5.4% of the Middle and 15% of the known Upper Palaeolithic individuals. Burial of the entire corpse was therefore seldom practised in

Palaeolithic times. Only most favoured individuals were buried.

Reflections on life and death began with late *Homo erectus* about 500,000–300,000 years ago. The great variety of mortuary practice and ritual in palaeolithic times is positive evidence of intensive reflections on life and death at that time. The majority of the mortuary practices and rituals for favoured dead were connected with defleshing and dismemberment of the corpse and fragmentation of bones of the deceased. Burial of the entire intact corpse, coexisting with other mortuary practices, reflects a new way of thinking by humans and a new form of confrontation with life and death during the daily life of Palaeolithic man. In contrast to other forms of mortuary practice connected with human interference in human corpses (defleshing, dismemberment and bone fragmentation), the entire and intact corpses carefully and intentionally buried at occupation sites have shown that it was most important for both the deceased and the living social group to preserve the entire and intact body of the dead. This is indicative of a very close connection between the deceased and the living community. Reflections on life and death in the Palaeolithic necessarily initiated reflections on the world in which the humans were living on in the afterworld. It has been suggested that Palaeolithic man was reflecting on an afterworld that was closely connected with the one in which he was living.

There was also clear evidence of cannibalism in Palaeolithic times, yet only practiced within mortuary ceremonies and at very few sites. Cannibalism cannot be considered a universal attribute of Palaeolithic man (Ullrich, 1989a; in press).

The results of investigations on bone modifications, mortuary practices, burial rites, skull-cult and cannibalism were considered in greater detail here, because they clearly demonstrated that disciplinary (i.e., anthropological) studies of bone modifications in connection with other disciplinary (i.e., archaeological) results (e.g., the archaeological context of fossil human bones) could not only raise a large number of interesting questions but they contributed to their solution too.

Fundamental studies have also been done on the problems of the hand-brain-tool dialectics in human evolution (Grimm, 1985) in order to understand the efficiency of selection in a system with an extra-organismic part (tool).

It should also be mentioned that a large number of papers on the history of human evolutionary research have been published in our country.

ARCHAEOLOGY

Archaeological research in eastern Germany has usually been connected with Palaeolithic sites, but there are also basic studies and analyses on the origins and evolution of labour and the labour process, their steps and effects on the evolution of mankind (Herrmann, 1983; 1984). Anthropogenesis started in his opinion with the onset of labour and was completed after the working process had established itself as the fundamental factor in the history of society. Starting from this definition of *Menschwerdung* — another one than mentioned above — four periods were distinguished in the evolutionary history of mankind. These are:

1. *Homo*-Deviation;
2. Anthroposociogenesis;
3. Socio-economic formation;
4. Development of ancient society.

The origins of modern humans and their culture were the subject of a detailed study, mainly under archaeological aspects (Struwe, 1989). An outline of the development of preparing and using artifacts and tools during the Palaeolithic was published by Feustel (1973). Weber & Schäfer (1983) have analysed numerous metric data of Palaeolithic stone tools in order to get detailed information on the function of these objects.

A textbook on the history of ancient society has been elaborated by archaeologists (Grünert, 1982).

ETHNOGRAPHY

There are only a few contributions from ethnography to human evolution, concerning especially the transition from Middle to Upper Palaeolithic and the development of the economy of Australian aborigines based on intensive field-work in Australia. (Rose, 1976; 1987 a, b).

ETHOLOGY

Ethologists in our country are very interested in the problems of human evolution. It was in 1974 that a first conference on the biological background of human behaviour was held in Cottbus (Johst, 1976). Since that time ethologists have very often discussed at conferences the problem of pre-human and human behaviour in close connection with the origins and evolution of man as well as — together with philosophers — the philosophical and ethical aspects of human behaviour and its relation with human society (Tembrock et al., 1978; Geißler & Hörz, 1988).

EVOLUTIONARY BIOLOGY

Evolutionary biologists have also focussed on the problem of the origins and evolution of Man. The evolution of biosocial structures (Stephan, 1977) and biological and philosophical aspects of animal and human societies (Löther, 1988) as well as biological foundations of the historicity of mankind (Freye, 1983) are only some topics.

PHILOSOPHY

Philosophical problems of the origins and evolution of Man and human society have been discussed in several papers (e.g., Bergner, 1982; Löther, 1988). A special investigation has focussed on the anthropogenesis and the materialistic dialectics and analysed in detail the philosophical-theoretical and methodological foundation of the anthropogenesis process (Foerster, 1981).

PSYCHOLOGY

Psychologists have analysed the development of mental processes and documented the specific form of human thinking, human language and human intelligence (Klix, 1980).

MULTIDISCIPLINARY AND INTERDISCIPLINARY CO-OPERATION

The multidisciplinary and interdisciplinary co-operation between the members of the working group "Menschwerdung" has been formed in a long and difficult process (Ullrich, 1987), but it was a necessary prerequisite for the start of the book project "Menschwerdung" and for its implementation.

The first step was characterized by a closer acquaintance made with the working methods of the different disciplines and with the expected contributions and possible results concerning the hominization process. This was achieved in colloquia and symposia where the facts and the general theoretical problems were discussed under different disciplinary aspects. Disciplinary thinking, misunderstandings and difficulties with different disciplinary terminologies had to be overcome in this first step. The process of multi- and interdisciplinary co-operation has been promoted by establishing a special group within the working group, responsible for preparing and writing the manuscripts for the book "Menschwerdung," and by discussing concrete questions of form and content of presenting the hominization process from multiple and interdisciplinary perspectives. It was necessary to find not only a uniform and common terminology, but also to combine various disciplinary facts and concepts and to view the unity of the logical and the historical considerations as the basis for new and deeper understanding of the hominization process. This process was a difficult and slow one. Some disciplines (e.g., anthropology, palaeontology, archaeology) were familiar with the evolutionary interpretation of facts (fossils, tools, etc.), whereas others (e.g., ethnography, ethology) were familiar with the analysis of recent phenomena and processes. Conclusions on earlier processes of hominid evolution had to be drawn from the co-existence of archaic and progressive structures in recent populations. But it was necessary to find a common basis. This second step of multi- and interdisciplinary co-operation started in 1980.

The third step of co-operation between different disciplines of natural sciences and social sciences was characterized by multi- and interdisciplinary discussions of the manuscripts for the book

"Menschwerdung." Twenty-one authors — all members of the working group, with the addition of a Czech archaeologist, were representing 11 disciplines: anthropology, archaeology, ethnography, ethology, evolutionary biology, geophysics, history, musicology, palaeontology, philosophy, psychology. They had written more than 50 papers. During two four-day closed meetings outside Berlin all of them were discussed by the members of the working group under multi- and interdisciplinary perspectives. Many, but not all differences and discrepancies in interpretation could be solved among the authors. After the revision of the manuscripts by the authors it was necessary to have the style standardized by one of the editors, since we did not want to have a book on the hominization process published with a collection of papers and without a uniform language. This work was completed by the end of 1986 and the manuscripts were handed to the publisher Akademie-Verlag in Berlin.

The book includes the following main chapters:

- biotic evolution and evolution of man;
- palaeogeography, palaeoclimate, floras and faunas in the Tertiary; dating methods; primate evolution; recent primates: distributions, habits;
- biosociology and environmental relations of primates; origins of language and thinking;
- early hominid evolution and biotic prerequisites for the origin of man; hypotheses on the origins of hominids; systematics and periodization of hominid evolution; the question of ramapithecines; evolution of primate brains; hand-brain-tool dialectics;
- Plio-Pleistocene environment in Africa; australopithecines; *Homo habilis*; labour — a new driving force in hominid evolution;
- origins of social life in early hominid evolution: palaeogeography, palaeoclimate, floras and faunas in Pleistocene; *Homo erectus* populations; Lower Palaeolithic archaeological cultures; *Homo erectus* from Bilzingsleben — his culture and environment; social life of *Homo erectus*
- origins of *Homo sapiens*; archaic *Homo sapiens*; archaeological cultures and way of life in the Middle Palaeolithic; labour, consciousness and language of archaic *Homo sapiens*;
- ancient society of hunter-gatherers at the end of the glacial period: *Homo sapiens* sapiens and his origins; origins of races; cultural development in the Upper Palaeolithic; art, cults and music of Upper Palaeolithic man; ethnographical aspects of hunter-gatherers;
- Man and the origins of Man in philosophical and anthropological concepts until the end of the 19th century; development of the anthopogenesis theory from Darwin to recent time; historical aspects of human evolutionary research; philosophical explanation

of the origins and evolution of Man in the 19th and 20th century;
- abbreviated catalogue of fossil hominid remains discovered until the end of 1990.

Unfortunately the publication of the book *Menschwerdung* (eds. J. Herrmann, H. Ullrich) was delayed until the end of August 1991. We had great trouble with the publisher. Meanwhile great political and economic changes occured in eastern Germany. The Akademie-Verlag refused to print our book, although the imprimatur was given by us. One reason for the refusal was that the book had been condensed under the impact of Marxist ideology of the German Democratic Republic. In our book *Menschwerdung* we had stressed, of course, the eminent significance of labour and the labour process in the evolution of humans and human society — but only because in heated interdisciplinary discussions we had recognized this significance, and not, because K. Marx and F. Engels more than one hundred years ago had also emphasized the significance of labour in the process of Menschwerdung. Their ideas on human evolution were more or less unknown to many scholars referring to human evolution. So they have been discussed in more detail in our book. The book *Menschwerdung* is therefore no Marxist textbook. It is, of course, a product of scientists from eastern Germany, written in the years up to 1986 and twice supplemented by new facts, new ideas and new literature until 1990. It is a product from eastern Germany with all the possibilities we had at our disposal, but first of all with the idea of a multidisciplinary and interdisciplinary approach to the origins and evolution of humans and human society.

In preparing this book *Menschwerdung* about 25 colloquia, symposia and other meetings have been organized by the working group discussing facts and theoretical aspects of the hominization process. The most difficult problem was to keep all the meetings of the working group so interesting, lively, efficient and successful that all the members, working in different institutes and museums in eastern Germany, kept their interest in co-operating for more than 12 years! In 1981 an international interdisciplinary conference "Anthroposoziogenese — biotischer und gesellschaftlicher Entwicklungsprozeß der Menschheit" was held in Weimar. The proceedings of this conference were published (Herrmann & Ullrich, 1985). Co-operation with the Department of Evolutionary Biology CSAV in Prague resulted in a common conference on the evolution of man in Jachymov (Czechoslovakia) 1980 — published in *Anthropologie* (Brno) 21, 1983. Members of our working group have also carried out intensive disciplinary human evolutionary research as mentioned above. Some of them had the possibility to visit important Plio-Pleistocene and/or Palaeolithic sites in Africa (Hadar in Ethiopia; Olduvai in Tanzania), Asia (Zhoukoudian in China) and Europe and to study human remains as well as archaeological and palaeontological finds. A bibliography entitled "Bibliographie zur

Menschwerdung und frühen Menschheitsentwicklung," compiled by order of the interdisciplinary working group, started with number 1 in 1979. Fourteen numbers with 7,000 titles of publications (since 1978) have been published (Ullrich, 1979-1990). This bibliography is available to all colleagues interested in it, by exchange of publications.

After finishing our book *Menschwerdung* we have, of course, recognized many unresolved problems of the origins and evolution of Man and human society. We recognised the problem of continuity and discontinuity and discussed it at an international interdisciplinary conference "Kontinuität und Diskontinutität in der Evolution des Menschen bis zur Herausbildung der Urgesellschaft" in Lutherstadt Wittenberg in 1988. The proceedings have been prepared for publication (Ullrich, in press).

There are also many articles in popular journals (e.g., *Urania* 1987-1988), published by the members of the working group, dealing with various aspects and problems of the origins and evolution of Man and human society. A considerable number of books have been published, too. Only a few topics can be mentioned here: human evolution (Donat and Ullrich, 1972; Herrmann, 1973; 1984; Ullrich, 1974; Feustel, 1976; Straaß, 1978); origins of the human family (Mohrig, 1980); woman in early history (Schlette, 1988); Palaeolithic sites Weimar-Ehringsdorf (Steiner, 1979) and Bilzingsleben (Mania & Dietzel, 1980); glacial period (Kahlke, 1981). A biography of Hauser, the Swiss archaeologist and discoverer of the Neandertal Man from Le Moustier (Drößler, 1988), was also published in eastern Germany.

FUTURE INTERDISCIPLINARY RESEARCH

The interdisciplinary working group "Menschwerdung" completed its work in 1989. Changes in the structuring of sciences in eastern Germany make it impossible to continue, yet they open up new possibilities for implementing the new Human Evolution International Interdisciplinary Project "Man and the Environment in the Palaeolithic." The aim of this project is to provide a broad interdisciplinary discussion and a broader interdisciplinary co-operation in the study of the origins and evolution of humans dependent on changes in the environment, and on the effects of the environment on humans and the effect of humans on the environment. The first symposium will be held in Neuwied probably in May next year. Further symposia will follow in conformity with fundings. In addition to this international interdisciplinary panel, the members of this project are going to prepare a book on "Man and the environment in the Palaeolithic" by the end of 1994. About 50 scientists representing a different spectrum of disciplines in most European countries, but also in USA, Canada and Israel, are interested in this new project. Project coordinators will be G. Bosinski (archaeologist), Neuwied, Y. Coppens (palaeontologist), Paris, and H. Ullrich (anthropologist, project organizer), Berlin.

The new project had to be organized under the auspices of the Humboldt-University in Berlin, because at the end of 1991, the Zentralinstitut für Alte Geschichte und Archäologie, where the working group "Menschwerdung" was established in 1977 and operated until 1989, will be suspended. The experiences gained in 12 years of multi- and interdisciplinary co-operation in the working group "Menschwerdung" will serve as a good starting platform for the project "Man and the environment in the Palaeolithic."

REFERENCES

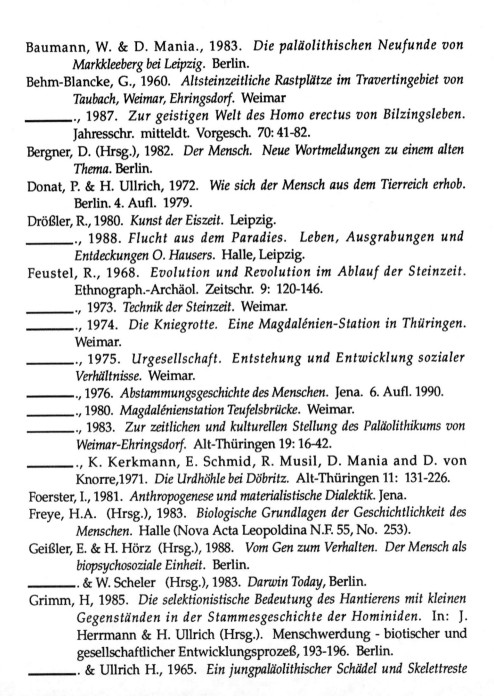

Baumann, W. & D. Mania., 1983. *Die paläolithischen Neufunde von Markkleeberg bei Leipzig.* Berlin.

Behm-Blancke, G., 1960. *Altsteinzeitliche Rastplätze im Travertingebiet von Taubach, Weimar, Ehringsdorf.* Weimar

————., 1987. *Zur geistigen Welt des Homo erectus von Bilzingsleben.* Jahresschr. mitteldt. Vorgesch. 70: 41-82.

Bergner, D. (Hrsg.), 1982. *Der Mensch. Neue Wortmeldungen zu einem alten Thema.* Berlin.

Donat, P. & H. Ullrich, 1972. *Wie sich der Mensch aus dem Tierreich erhob.* Berlin. 4. Aufl. 1979.

Drößler, R., 1980. *Kunst der Eiszeit.* Leipzig.

————., 1988. *Flucht aus dem Paradies. Leben, Ausgrabungen und Entdeckungen O. Hausers.* Halle, Leipzig.

Feustel, R., 1968. *Evolution und Revolution im Ablauf der Steinzeit.* Ethnograph.-Archäol. Zeitschr. 9: 120-146.

————., 1973. *Technik der Steinzeit.* Weimar.

————., 1974. *Die Kniegrotte. Eine Magdalénien-Station in Thüringen.* Weimar.

————., 1975. *Urgesellschaft. Entstehung und Entwicklung sozialer Verhältnisse.* Weimar.

————., 1976. *Abstammungsgeschichte des Menschen.* Jena. 6. Aufl. 1990.

————., 1980. *Magdalénienstation Teufelsbrücke.* Weimar.

————., 1983. *Zur zeitlichen und kulturellen Stellung des Paläolithikums von Weimar-Ehringsdorf.* Alt-Thüringen 19: 16-42.

————., K. Kerkmann, E. Schmid, R. Musil, D. Mania and D. von Knorre,1971. *Die Urdhöhle bei Döbritz.* Alt-Thüringen 11: 131-226.

Foerster, I., 1981. *Anthropogenese und materialistische Dialektik.* Jena.

Freye, H.A. (Hrsg.), 1983. *Biologische Grundlagen der Geschichtlichkeit des Menschen.* Halle (Nova Acta Leopoldina N.F. 55, No. 253).

Geißler, E. & H. Hörz (Hrsg.), 1988. *Vom Gen zum Verhalten. Der Mensch als biopsychosoziale Einheit.* Berlin.

————. & W. Scheler (Hrsg.), 1983. *Darwin Today,* Berlin.

Grimm, H, 1985. *Die selektionistische Bedeutung des Hantierens mit kleinen Gegenständen in der Stammesgeschichte der Hominiden.* In: J. Herrmann & H. Ullrich (Hrsg.). Menschwerdung - biotischer und gesellschaftlicher Entwicklungsprozeß, 193-196. Berlin.

————. & Ullrich H., 1965. *Ein jungpaläolithischer Schädel und Skelettreste aus Döbritz, Kr. Pößneck.* Alt-Thüringen 7: 50-89.

Grünert, H. (Hrsg.), 1982. *Geschichte der Urgesellschaft.* Berlin.

Herrmann, J., 1973. *Die Entwicklung der Menschheit.* Berlin.

———., 1983. Tool making and first stages of labour-factors and the results of human evolution. *Anthropologie* (Brno) 21: 7-17.

———., 1984. *Die Menschwerdung. Zum Ursprung des Menschen und der menschlichen Gesellscchaft.* Berlin. 4. Aufl. 1988.

———. (Hrsg.), 1989. *Archäologie in der Deutschen Demokratischen Republik. Denkmale und Funde.* Leipzig, Jena, Berlin.

———. & Ullrich H. (Hrsg.), 1985. *Menschwerdung - biotischer und gesellschaftlicher Entwicklungsprozeß.* Berlin.

———. & Ullrich H. (Hrsg.), 1991. *Menschwerdung. Millionen Jahre Menschheitsentwicklung - natur - und geisteswissenschaftliche Ergebnisse. Eine Gesamtdarstellung.* Berlin.

Jäger, K.D. & Heinrich W.D., 1982. The travertine at Weimar-Ehringsdorf - an interglacial site of Saalian age? In: Quaternary glaciation in the western hemisphere (International geological correlation programme, project 73/1/24). Report No. 7: 98-113. Prague.

Johst, V. (Hrsg.), 1976. *Biologische Verhaltensforschung am Menschen.* Berlin. 2. Aufl. 1982.

Kahlke, H.D. (Hrsg.), 1974. *Das Pleistozän von Weimar-Ehringsdorf. Teil I.* Berlin.

———. (Hrsg.), 1975. *Das Pleistozän von Weimar-Ehringsdorf. Teil II.* Berlin.

———. (Hrsg.), 1977. *Das Pleistozän von Taubach bei Weimar.* Berlin (Quartärpaläontologie 2).

———. (Hrsg.), 1978. *Das Pleistozän von Burgtonna in Thüringen.* Berlin (Quartärpaläontologie 3).

———., 1981. *Das Eiszeitalter.* Leipzig, Jena, Berlin.

———. (Hrsg.), 1984. *Das Pleistozän von Weimar - Die Travertine im Stadtgebiet.* Berlin (Quartärpaläontologie 5).

Klix, F., 1980. *Erwachendes Denken. Eine Entwicklungsgeschichte der menschlichen Intelligenz.* Berlin. 2. Aufl. 1983.

Löther, R. (Hrsg.), 1988. *Tiersozietäten und Menschengesellschaften. Philosophische und evolutionsbiologische Aspekte der Soziogenese.* Jena.

Mai, D.H., D. Mania, T. Nötzold, V. Toepfer, E. Vlček & W.D. Heinrich, 1983. *Bilzingsleben II. Homo erectus — seine Kultur und seine Umwelt.* Berlin.

Mania, D. 1990. *Auf den Spuren des Urmenschen. Die Funde von Bilzingsleben.* Berlin.

———. & A. Dietzel, 1980. *Begegnung mit dem Urmenschen. Die Funde von Bilzingsleben.* Leipzig, Jena, Berlin.

———. & U. Mania, 1988. Deliberate engravings on bone artefacts of Homo erectus. *Rock Art Res.* 5, 2: 91-132.

———. & V. Toepfer, 1973. Königsaue. *Gliederung, Ökologie und mittelpaläolithische Funde der letzten Eiszeit*. Berlin.

―――., V. Toepfer. & E. Vlcek, 1980. *Bilzingsleben I. Homo erectus – seine Kultur und seine Umwelt*. Berlin.

―――. & E. Vlček, 1987. Homo erectus from Bilzingsleben (GDR) - his culture and his environment. *Anthropologie* (Brno) 25: 1-45.

―――. & T. Weber, 1986. *Bilzingsleben III. Homo erectus - seine Kultur und seine Umwelt*. Berlin.

Mohrig, W., 1980. *Wie kam der Mensch zur Familie?* Leipzig, Jena, Berlin.

Musil, R., 1985. *Die Fauna der Magdalénien-Siedlung Oelknitz*. Weimar.

Rose, F., 1976. *Australien und seine Ureinwohner*. Berlin.

Rose, F.G.G., 1987a. *The Traditional Mode of Production of the Australian Aborigines*. Sydney.

―――., 1987b. Der Übergang vom Mittel – zum Jungpaläolithikum vom Standpunkt eines Ethnographen. *Ethnograph.-Archäol. Zeitschr.* 28: 185-208.

Schäfer, D., 1988. *Merkmalsanalyse mittelpaläolithischer Steinartefakte*. Unveröff. Diss. Humboldt-Universität Berlin.

Schlette, F. (Hrsg.), 1980. *Die Entstehung des Menschen und der menschlichen Gesellschaft*. Berlin.

―――., 1988. *Von Lucy bis Kleopatra. Die Frau in der frühen Geschichte*. Berlin.

Steiner, W., 1979. *Der Travertin von Ehringsdorf und seine Fossilien*. Wittenberg.

Stephan, B., 1977. *Die Evolution der Sozialstrukturen*. Berlin.

Straaß, G., 1978. *Rassen - Herkunft und Zukunft. Urteile und Vorurteile*. Berlin.

―――., 1984. *Der Mensch - Krone der Evolution*. Berlin.

Struwe, R., 1989. *Die Herausbildung des Jetztmenschen und seiner Kultur*. Unveröff. Diss. Humboldt-Universität Berlin.

Tembrock, G., E. Geißler & W. Scheler W. (Hrsg.), 1978. *Philosophische und ethische Probleme der modernen Verhaltensforschung*. Berlin.

Ullrich, H., 1974. *An der Schwelle der Menschheit*. Leipzig, Jena, Berlin.

―――., 1975. Bemerkungen zu den Fundumstäden und zur Deutung der menschlichen Skelettreste aus der Urdhöhle bei Döbritz. *Zeitschr. Archäol.* 9: 307-318.

―――., 1978. Kannibalismus und Leichenzerstückelung beim Neandertaler von Krapina. In: M. Malez (red.). *Krapinski pracovjek i evolucija hominida*, 293-318. Zagreb.

―――., 1979a. Artifizielle Veränderungen am jungpaläolithischen Schädel von Cioclovina (SR Rumänien). *Annuaire Roumain d'Anthrop.* 16: 3-12.

―――., 1979b. Artifizielle Veränderungen am Occipitale von Vértesszöllös. *Anthrop. Közlem.* 23: 3-10.

―――., 1979-1990. *Bibliographie zur Menschwerdung und frühe Menschheitsentwicklung*. No. 1-14. Berlin.

————., 1982. Artificial injuries on fossil human bones and the problem of cannibalism, skull-cult and burial rites. *Anthropos* (Brno) 21: 253-262.

————., 1984. Petralona - eine rituelle Schädelbestattung? *Ethnograph. Archäol. Zeitschr.* 25: 585-627.

————., 1986. Manipulations on human corpses, mortuary practice and burial rites in Palaeolithic times. *Anthropos* (Brno) 23: 227-236.

————., 1987. "Menschwerdung" — ein interdisziplinäres Projekt. *Spectrum* 18, 2: 5-7.

————., 1989a. Kannibalismus im Paläolithikum. In: F. Schlette & D. Kaufmann (Hrsg.). *Religion und Kult in ur und frühgeschichtlicher Zeit*, 51-71. Berlin.

————., 1989b. Neandertal remains from Krapina and Vindija - mortuary practice, burials or cannibalism? In: O.G. Eiben (ed.), *European Populations in Past, Present and Future*, 15-19. Budapest.

————., 1990a. Zu einigen Grundlinien der Geschichte der Anthropogeneseforshung. In: S. Kirschke (Hrsg.). *Grundlinien der Geschichte der biologischen Anthropologie*, 77 - 90. Halle.

————., 1990b. German Democratic Republic. In: R. Orban (ed.). *Hominid remains. An up-date.* No. 3 (in press).

————., 1991. Totenriten und Bestattung im Paläolithikum. In: F. Horst & H. Keiling (Hrsg.). *Bestattungswesen und Totenkult in ur und frühgenschichtlicher Zeit*, 23-34. Berlin.

————., (Hrsg.), (in press). *Evolution des Menschen. Kontinuitäten und Diskontinuitäten.* Neuwied.

————., (in press). Kontinuität Wandel in den Totenriten und Jenseitsvorstellungen des Paläolithischen Menschen. In: H. Ullrich (Hrsg.), *Evolution des Menschen. Kontinuitäten und Diskontinuitäten.* Neuied.

————., (in press). Modifications of fossil human bones: current status of facts and interpretations. *Anthropologie* (Brno.).

Ullrich H., (in press). Palaeolithic burials - an anthropological approach. Proceedings of the Third Symposium on Upper Palaeolithic, Mesolithic and Neolithic populations of Europe and the Mediterranean Basin. Budapest.

————. (Ms.). *Kannibalismus, Schädelkult und Bestattung in paläolithischer Zeit.*

Vlček, E., 1978. A new discovery of Homo erectus in Central Europe. *Journ. Human Evol.* 7: 239-251.

————., 1985. Der fossile Mensch von Weimar-Ehringsdorf. In: J. Herrmann & H. Ullrich (Hrsg.). *Menschwerdung - Biotischer und Gesellschaftlicher Entwicklungsprozeß*, 111-117. Berlin.

Weber, T. & Schäfer D., 1983. Analytische Betrachtungen und historische Interpretation altpaläolithischer Artefaktkomplexe. *Zeitschr. Archäol.* 17: 1-30.

CHAPTER 9

HUMAN EVOLUTION IN ALBANIA FOR THE QUATERNARY PERIOD

A.B. FISTANI

Until recently in Albania no human fossil remains dating to the Pleistocene had been found. Prior to 1950, even the idea of Albanian research into problems of human evolution was questionable. This was changed when palaeolithic industries were discovered in Northern Albania. In fact this paper is focussed principally on three palaeolithic sites discovered during these last 10 years south-east of Shkoder (Fig. 1). This material, representing lithic sequences discovered in both open air sites and in caves, shows the presence of early humans and is encouraging in the possibility of tracing fossil remains.

A retrospective search of past work done in our country yields limited material. Our background is weak. Many scholars have visited and explored Albania, especially during the last hundred years, but none of them reported or even supposed the theoretical possibility of palaeolithic traces left by primitive groups. Thus the stone age in this region of the Balkan peninsula was quite unknown. Some years after, not far from the actual Albanian border, however, the early discovery of this kind of evidence which seems to belong to the Lower Palaeolithic, was made by the French researcher E. Patte, (1918), near Bitolje at Monastir.

In 1938 the palaeontologist, B. von Richtoffen, (1939), found two single flakes discovered near Dajti at Tirana, and concluded that Ice Age Palaeolithic Man had lived here. This modest beginning was followed by an important discovery, made by Italian archaeologists in the campaigns of 1938-1939, in the southern parts of the Adriatic shore, in the village of Xara close to Butrinti.

Although there were great expectations of further exploration of these palaeolithic traces in this region of the Balkan peninsula, no further work has been done at these sites. Unfortunately they have been left as intact as the day of their first discovery.

Historically speaking, the search for the first traces of palaeolithic populations in Albania began in 1938. In spite of their modest contribution they offer only a kind of hint of the real future possibilities in the search for data about human evolution. On the other hand, it seems quite crucial now to undertake surveys in this region which occupies a central geographical position between the South and the North of the peninsula.

The discovery of the Gajtan site in 1981, opened a new state in human palaeontological research and Pleistocene palaeontology. Gajtan I with its early primitive dwelling floor extends the evidence of the first Palaeolithic Civilization in Albania to a larger geological scale, going back to the ancient period of the Middle Pleistocene. The human palaeontology of this period was quite unknown before. Thus the elongation of the stratigraphical sequence of the Quaternary in our country was very important. At the same time the palaeontological data, where the faunal assemblages are found in association and in the same context with the lithic assemblages, together form the first evidence that sheds light on what may be deduced as human evolution in our country. Of course, we are conscious that with a lack of fossil human remains dating to the Pleistocene, it is difficult to attempt a direct correlation between palaeolithic industries and human evolution. Strictly speaking, in this case, we have to make do with cultural evidence of human groups who left their traces as tool-makers, but whose physical form is still unknown for us. It is a fact, that many authors in different countries in the same situation (a total lack of proven fossil hominids) are obliged to follow the same methodological approach by concentrating their analyses on cultural evolution. In this field the extrapolations which aim to correlate cultural evolution with the physical evolution of prehistoric Man are almost hypothetical and such territory ought to be undertaken with caution.

One other important argument for the presence of prehistoric Man in Albania is that of the geomorphology of the Quaternary period. Studies on the presence of different traces left by glacial periods in the Albanian high mountains also create questions. The most important early glaciations are discussed by Bourcart (1920) and Cvijic (1901), who attest to the presence of these periods in the Albanian Alps only.

On the other hand, this area, geographically speaking, is interlaced by high mountains associated with fertile valleys and numerous hills which lie near the flat widespread lands just close to the Adriatic sea shore. As a matter of fact, the weather today continues to be Mediterranean and frequently warm and wet, and was likely hospitable to different primitive civilizations during the Pleistocene period.

Calcareous massifs likewise are both numerous and important, particularly along the Eastern border of the ridge of mountains of Albania, and there are many karstik fissures, caves and canyons situated near the water sources or rivers. Also the abundant presence of different kinds of hard stones such as the silex, quartz, flintstone, etc., provides the necessary raw materials for exploitation by earlier prehistoric peoples. In fact, the majority of different areas capable of supporting hominid existence during the Quaternary period and able to yield such evidence are as yet totally unexplored from this point of view.

FIGURE 1

THE SOUTH-EAST AREA OF SHKODER WITH FIVE GENERAL SITES (BLACK DOTS) WHERE THE DESCRIBED PALAEOLITHIC SITES WERE DISCOVERED (1981-1991)

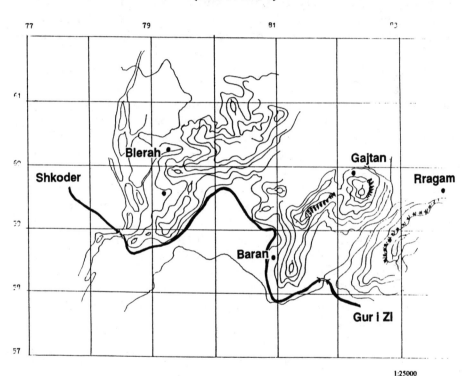

1:25000

If we consider our country in terms of the political borders, it actually represents a small geographical area. But this ancient ground is an inseparable part of the Balkan peninsula which lies in a central region bordering the land to the South (Greece) and to the North (Yugoslavia) just near the Adriatic and Ionic seas. It is necessary to emphasize that our country, geologically speaking is really of the same constitution as other parts of this peninsula, and has similar floral and faunal characters. The discoveries of fossil remains in limitrophic areas attest to the idea that this geographical region, including our country, contributed in the process of Mankind. Undoubtedly human evolution took place also in this Mediterranean area. From this prospective, as research intensifies, the possibility for future discoveries of human remains of Pleistocene age exists in Albania too.

When the Laboratory of Human Palaeontology found evidence of primitive man in Albania, two kinds of research were undertaken to investigate this Stone Age Man. Although there has been reduced support from scientific institutions for extending this new research, systematic exploration of the caves of Gajtan I and II has been carried out over the last ten years.[1] Secondly, the exploration of open-air surface sites such as river valleys or terraces has been fruitful in yielding rich material, especially southeast of Shkoder. In addition the collection of occasional fossil mammals found during mining activities or different building works provides good results, as is the case of the quarry of Shahinove at Berati, and Bezhan in Korçe. Even with all this, it is a fact that, until now, the entire geographical distribution of Palaeolithic Cultures in our country is not well known. No general survey has been made.

Although both foreign and Albanian research has been undertaken including many archaeological projects, few of them have published, or reported finding palaeolithic tools. It is a pity that no mammal remains were conserved totally, or studied, for any earlier archaeological excavation. In general the Palaeolithic research in Albania was both undervalued and regarded with a great deal of scepticism by some specialists.[2] None of them believed in the possibility that our territory might contain large mammals or fossil human remains, until the first site was reported by the Laboratory of Human Palaeontology at Shkoder.

It is an irrefutable fact that human evolution and Darwinian theory were studied in middle and high schools of Albania, but no research program applying this knowledge in this field was undertaken for many years. Human evolution now is quite a new concept recently adopted only by the program of our research, and is based on the preliminary lithic plus faunal evidence discovered during the last 12 years of work. This program is new and many problems remain to be solved in the future. Almost the whole of Albania, from this point of view, is unexplored. Before our research, only archaeological finds which began from the Neolithic and extended to the Middle Ages were brought to light. Only after 1981 were some caves explored by archaeologists, but neither Palaeolithic or Mesolithic layers were reported.

Based on bibliographical studies in Anthropology, I will now present in a descriptive and chronological way the most important studies and the methodological aspects followed during this last century in Albania. This short review, in fact, is outside of our area of activity, but to encompass a proper sketch of Albanian research in human evolution it seems necessary to discuss also these post-Pleistocene studies.

In an anthropological sense, since Boué (1840) many scholars paid attention to the physical aspect of the Albania people. Naturally nowadays these papers seem to be descriptive observations. Later, many other anthropologists wrote more advanced studies on the same topics. In fact, German or French authors made many acute scientific observations in the field of Ethnic Anthropology about the Albanian race. Chronologically, the most important studies are those of Zampa (1886) who studied the cephalic index of some Albanian persons from Cosenza (Italy) and afterwards studied some skulls from Shkoder. A single skull from North Albania was described by Wirchoff (1877), and later, Hamy (1900) described 8 skulls from Shkodra. Glüeck (1897) studied nine Albanian skulls recovered at Delbinisht.

A number of eminent anthropologists and scholars including Gobineau (1884), Haberlandt and Labzelter (1919), Puttard (1921) and Weininger (1934) published studies about the racial characteristics of Albanians.

In 1972, during the Congress of Illyrian Studies held at Tirana, there was general agreement that it was urgent that Albanian anthropological science should be supported. However, this was based on the concept that such work would be directed toward the notion of tracing a single ethnic background for the Illyrian people. There were many reasons underlying this focus, concentrated on the study of the Albanian race and its origins. This approach was influenced by a political and demographic fact. Although the Albanian people may be scattered throughout the world, and the Albanian territory partitioned in earlier times, only the Albanian people including Kosova speak the Albanian language. Secondly, a major influence was the great number of earlier studies made by eminent foreign scientists who originated this line of enquiry.

Such studies, of course, are limited to the period of time encompassed. Although they make a precious contribution about the origin of Albanians from Illyrians, they do not add anything to our understanding of the Quaternary period of human evolution.[3]

LOWER PALAEOLITHIC AGE

In Albania, for many years, the discovery of Palaeolithic in general and especially Lower Palaeolithic traces was considered nearly taboo. Any theoretical attempt in this sense was met with disbelief primarily by archaeologists who worked in this field. Many theoretical opinions were based on the erroneous idea that this geographical area was one of deprivation and could not contain such vestiges of Palaeolithic Cultures. If any kind of Palaeolithic traces did exist here, it was believed that they could only be from the Upper Palaeolithic, Aurignacian Culture. This general thought was influenced perhaps by the extreme lack of research and by the general opinion that such Cultures (Aurignacian) derived from the oriental areas.

All the same, some prominent authors of the recent past (Reshetov, 1966), believed that because of the extreme isolation of the Balkan peninsula, prehistoric Man could not have evolved either physically or culturally, in the Pleistocene of these areas.

But in 1981 the discovery of extinct faunal elements — fauna of *Macaca, Dicerorhinus, Ursus thibetanus,* etc. — associated with crude lithic implements provided a reason to justify extending our research. The program established at the Laboratory of Human Palaeontology has resulted in the discovery of abundant Palaeolithic industries. They represent different evolutionary stages in the area south-east of Shkoder, including Gajtan I and II, Baran which is an open-air site, and others discovered recently, which belong to the Upper Palaeolithic, for example Bleran, another open-air site near the river Kiri. Following the discoveries made recently southeast of Shkodra, we have adopted studies and methodologies in line with other discoveries in the sciences, rather than following the old and traditional archaeological methods of earlier days.

BARAN (RENCE) (FIG., 2, 3, 4, 5, 6)

This open-air site situated nearly 0.8 km from the Gajtan site, extends along the western face of the eponyme hill. This important Palaeolithic site, which was discovered only in 1985,[4] was brought to light by chance during the opening up of new lands. A rich artifactual sequence of lithic tools, principally in jasper, was recovered over many years of field work, but neither mammalian bones nor human fossil remains were found. The soil is acidic.

Many crude components, such as hand-axes, bifaces, choppers, chopping-tools, cores and archaic crude large flakes were brought to light, chiefly in the second terrace. The occurrence of such archaic elements is more pronounced in this second terrace; nevertheless the distribution of tools was widespread and occupied a large surface of the entire hill.

It is important to underline that nearly all these elements retain large negatives, i.e., a large flaking technique was used, similar to the Acheulian-Abbevillian technique of employing a stone hammer to prepare the tools.

The artifactual flake sequence is typically crude and not specialized, although the stone of Baran is concoidal. Over 60% of the flakes contain a cortical heel associated with an angle of nearly 90°, or even more pronounced. This angle occurs between the outer face of the flake and the heel, giving the impression of a Clactonian type of flaking. Most of the flakes are extremely large, and demonstrate close affinity with the proto-

Levallois homologues discovered in Europe. Among the large number of flakes in the assemblage, only a small number are true Levallois flakes, which means that this technique was used very rarely and was little known. On the other hand, the typological categories of many other tools are not fixed, and a lot of artifacts are intermediate between other forms. Furthermore, a large number of elements were produced on crude quartzite blocks.

FIGURE 2 - BARAN (RENCË)
TWO TYPICAL BIFACES DISCOVERED IN THIS OPEN-AIR SITE

THE LEFT ONE IS FLAKED WITH LARGE FLAKING TECHNIQUE (ABEVELLIAN TECHNIQUE). THE RIGHT IS SIMILAR TO THOSE DISCOVERED IN DIFFERENT FRENCH SITES OF LOWER PALAEOLITHIC.

FIGURE 3
ARCHAIC IMPLEMENTS ON FLAKES
BARAN (RËNCE)

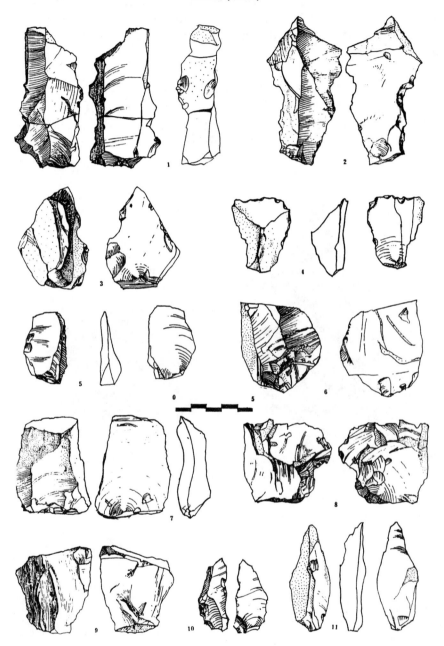

FIGURE 4
ARCHAIC IMPLEMENTS, LARGE BLADE-FLAKES
BARAN (RËNCE)

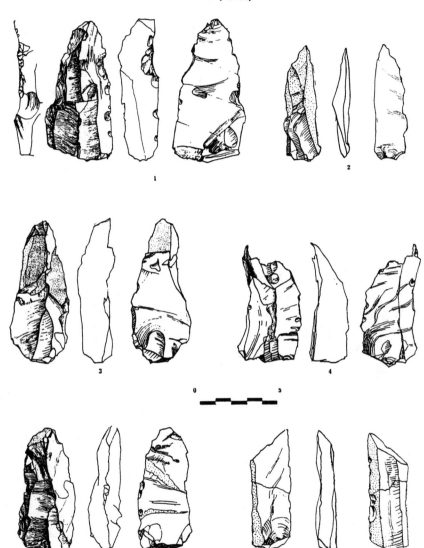

FIGURE 5
ARCHAIC IMPLEMENTS, LARGE CRUDE FLAKES
BARAN (RËNCE)

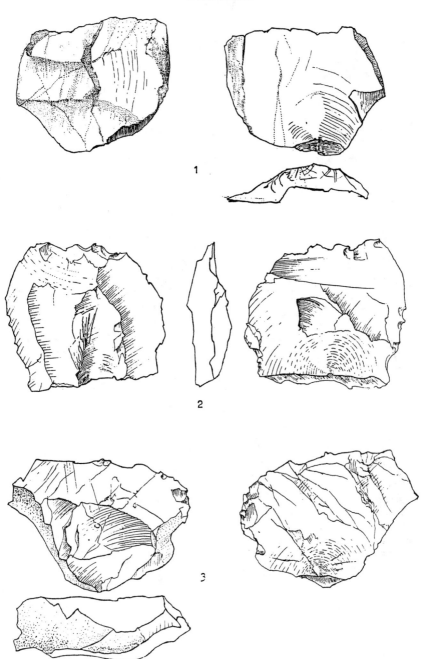

FIGURE 6
THE FUNCTIONS BETWEEN WIDTH OF FLAKES AND HEEL WIDTH IN SOME BARAN INDUSTRY-FLAKES

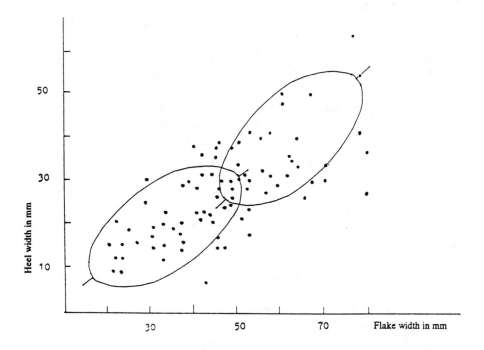

In these samples a small number of elements, such as flakes or discoidals, seem to show both a typological and technological evolutionary

In these samples a small number of elements, such as flakes or discoidals, seem to show both a typological and technological evolutionary sequence. The overall characteristics of the Baran industry suggest that they represent most probably a mixture from two Palaeolithic occupations of the same area, using the same raw material and living temporarily in the open-air, as hunters. Thus, we are dealing with the same hominine lineage, which perhaps evolved at least technologically in a limited area, or *in situ*, for a short span of time.

Based mainly on typological observation, tentatively we propose that the Baran industry is of pre-Mindel age in the Quaternary period. In any case, attempting to assign a cultural association with this original industry produces some difficulties; in fact it is not easy to characterize this stone age industry culturally. In this part of the Balkan peninsula where the paucity of prehistoric sites is pronounced, the cultural designation must be considered tentative. For this reason and in order to respect recent anthropological rules, we have tried to define not its cultural affinities, but to characterize it by comparing it with similar sites discovered elsewhere in Europe.

In fact several points of similarity can be observed between the Baran industry and those of other open-air sites discovered in Catalonia, Spain in the oldest sites such as in Montgri and La Selva, reported by Josep Canal i Roquet and Narcis Soler i Masferrer, (1976). These industries demonstrate many typological affinities for many of the Baran lithic elements. In general the lack of Levallois technique, the presence of very large flakes, the presence of pics (pointed heavy implements), the use of crude raw material, the lack of a fixed typological style on different elements are quite common in both cases in spite of their great geographical distance. Although the material is not identical, we can discern many analogies between these two industries developed during the Lower Palaeolithic in open-air sites. Some other elements such as choppers, discoids, large flakes or technological indices show a particularly striking similarity between these two sequences. Resemblances also occur for isolated single elements, such as flakes of the Baran and those of Djarkoutan industry discovered in Russia by Ranov (Gabori, 1976).

It seems evident that the Baran industry includes many archaic elements, although such similarities in this style are widespread throughout Europe. During the Pleistocene, such industries are associated with the first expansions of *Homo erectus*, which offers an enormous time span in which to place them — from the early Pleistocene to the last part of the Middle Pleistocene. Reasonably, Bonifay (1989) termed and characterized this little known period, especially for Europe, as *Trés Ancien Paléolithique*.

Seen from this perspective, two components of the Baran industry can be discerned by their location on the site: (1) a large number of flakes which show a more primitive development in the relationship between the width of the tool and its heel (Fig. 3, 4, 5, and 6) — this material is found in or near one specific area in terrace II; and (2) large-sized elements which were found in terrace II and can be termed as heavy duty implements. At the same time these two groups of tools are divided typologically and by their size.

In total, all primitive as well as more advanced elements display an inferior index of Levallois technique II. Also the first sequence of Archaic flakes and cores (Fig. 3, 4, 5) shows a complete absence of Levallois technology. The second sequence, more evolved, demonstrates a slight advance in this sense, and it appears to be a derivative of the first artifactual stage, representing perhaps a later phase of the Baran industry. Many typological and technological features are present, which might correspond to the same traditional method of flaking the stone tools.

In anticipation both of other studies and other discoveries in areas near Baran, we argue that the sequences of stone tools from this site belong to an evidently original Stone Age Culture of the Lower Palaeolithic, which deserves to be termed by its local name, the *Baran Industry of the Lower Palaeolithic*.

GAJTAN SITE (FIG. 7, 8, 9, 10, 11 AND 12)

In fact the real justification, although meagre, for beginning to study human evolution in our country, came from the discovery of the Baran and Gajtan sites. Gajtan I is a primary site with stratigraphical sequences containing principally two distinctive cultural levels, the lowermost (Fig. 7) with Acheulian features and the uppermost probably representing Mousterian activity (Fig. 8 to 11). Thus the Gajtan site, which is a semi-rock shelter, was occupied during the Middle Pleistocene by prehistoric Man. Two evident cultural levels which yielded both faunal and lithic artifactual elements are present. The lowermost level, the richest one, yielded a primitive artifactual sequence. Animal remains are distributed throughout this cultural stratum. But the evidence from the mode of distribution, both of lithic and of faunal material, in the two levels differs. Consequently this situation does not allow us an opportunity to follow the trends of their cultural evolution internally within Gajtan I deposits themselves. The great difference in time between the different layers of Gajtan I deposits impels us to find analogies in relation to other sites that are chronologically similar. One might think that those of Baran for the lowermost levels would be an obvious comparison, but only a few similarities were encountered among implements of Gajtan and this later open-air site.

The lithic industry of Gajtan I is almost entirely independent of the Baran lithic sequences, and demonstrates many characteristics which make it close to those considered to belong to the *Trés Ancien Paléolithique* of Europe. In the lowermost levels of Gajtan, the pebble tools (Fig. 11-3) are archaic and one can deduce that little care was taken by tool-makers with their preparation. A pronounced absence of flake implements, perhaps caused by their use of a raw material with poor conchoidal characteristics, a primitive technique of flaking the implements, the presence of some atypical bifaces or proto-bifaces, the absence of standardization of all tools, the fact that 40 % of the tools were choppers or chopping tools, a drastic lack of the Levallois technique, and the use of different kinds of hard raw material: all these are some characteristics of this sequence in Gajtan I.

In contrast, the uppermost layers of the Gajtan deposits seem almost a poor sequence, consisting mainly of flakes, but evidentl more evolved, both typologically and technically (Fig. 8).

In general, the faunal remains of this assemblage correspond to the restricted spectrum of animals associated with subsistence hunting. The lowermost level yielded a type of bone industry, which seems to have been used by primitive dwellers of this site. These fragmented bone remains show traces of human activity. These, along with antlers and bear skulls, were recovered near the main opening to the site. Some *Capreolus* skulls, intentionally fragmented, and browtines of shed antlers of red deer *Cervus elephus* display beam and browtines fragmented at the same distance from their base (Fig. 12), suggesting human use of this material. The presence of bone remains in this site is perhaps also due partly to the role played by carnivores, but only to a slight degree. The entire absence of articulated skeletal elements, the frequency of bone fragments, the lack of the so-called "charriage à sec" are important taphonomic signs suggesting the human origin of this material. The use of bone material adds a new dimension to our understanding of this prehistoric culture.

The available evidence suggests that the faunal remains from Gajtan I are essentially a Middle Pleisocene fauna, which contain some ancient elements surviving until this epoch,[3] such as *Ursus thibetanus, Macaca sylvana pliocena* or *Dicerorhinus cf. merckii.* The presence of this rich faunal material provides to a chronology, and gives us some indication of the regional palaeoecology. The presence of Red deer and Bison suggests open grasslands, whereas the presence of *Capreolus* and *Macaca* suggests the presence of forests. The numerous cervid remains are a valuable indicator of a climate characterized by moist and warm weather, which occurred during this part of the Middle Pleistocene.

Furthermore, Gajtan is a primary site with a chronological sequence built on undisturbed sediments with two principal cultural levels. The principal occupation was by an early Palaeolithic group contemporaneous with characteristic Middle Pleistocene faunal remains such as *Macaca, Dicerorhinus, Ursus thibetanus, Ursus cf. deningeri, Canis lupus mosbachensis, Capreous capreolus, Bison and Tortoise,* representing the hot-moist weather of the Holstein interglacial.

MIDDLE PALAEOLITHIC AGE (FIG. 13)

The first evidence of Middle Palaeolithic occupation in Albania was reported by Richtoffen (1938). In North Albania such studies did not occur until 1978-1985 with the first surface recovery of tools — a dozen typical Mousterian tools — from the valley of Gajtan near the eponyme site (Gajtan).

FIGURE 7
GAJTAN I SITE - STRATIGRAPHIC SEQUENCES

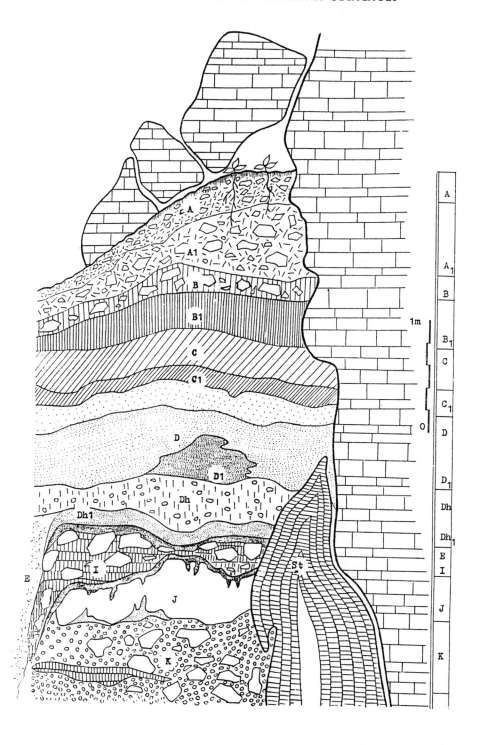

FIGURE 8
GAJTAN I SITE - IMPLEMENTS OF MOUSTERIAN OUTLINE
RECOVERED IN UPPERMOST LEVELS.
1, 2, AND 4, 5 ARE DISCOVERED IN LAYERS C-C$_1$;
THE IMPLEMENTS 3, 6, 7, 8 AND 9 ARE DISCOVERED IN LAYER D

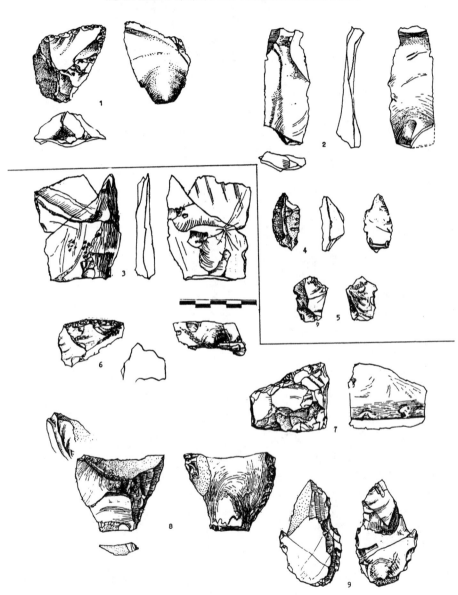

FIGURE 9
GAJTAN I SITE - ARCHAIC IMPLEMENTS OF AUHEULIAN OUTLINE
RECOVERED IN THE LOWERMOST LEVELS
LIVING FLOOR - LAYER DH, DH₁, J, K

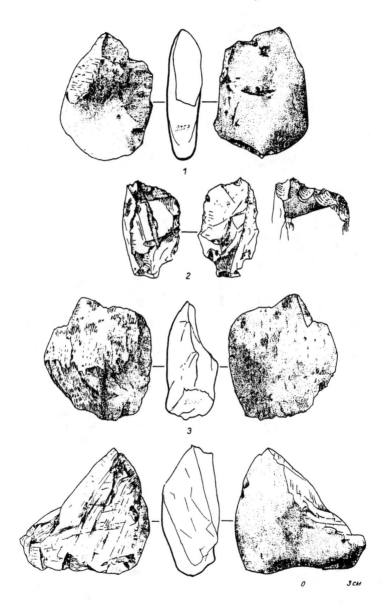

0   3 CM

FIGURE 10
GAJTAN I SITE - ARCHAIC IMPLEMENTS (CHOPPERS)
FROM LOWERMOST LEVELS

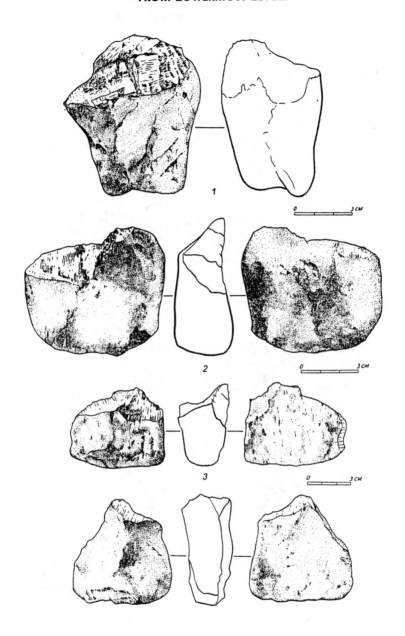

FIGURE 11
GAJTAN I SITE -ARCHAIC IMPLEMENTS, BIFACES AND PEBBLE TOOLS FROM THE LOWERMOST LEVELS

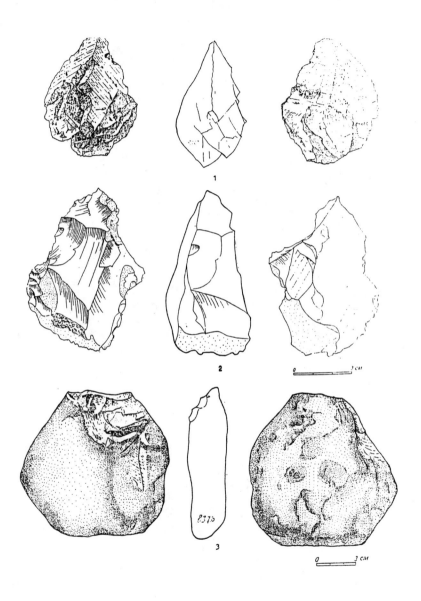

Figure 12
Gajtan I Site - Bones Conserving the Traces of Human Intervention (Antlers of Red Deer and Skulls of Roe Deer)
Fragmented Intentionally

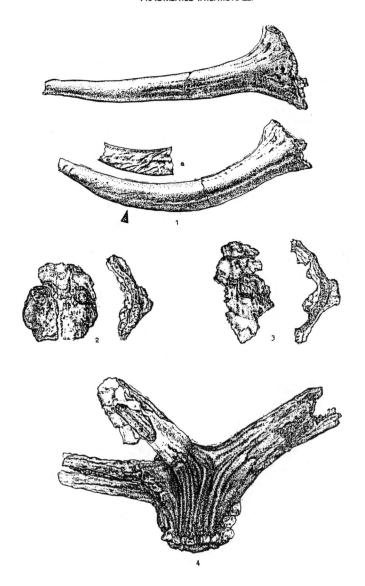

FIGURE 13
GAJTAN VALLEY - SOME TYPICAL MOUSTERIAN IMPLEMENTS RECOVERED IN VICINITY AREAS OF GAJTAN I SITE

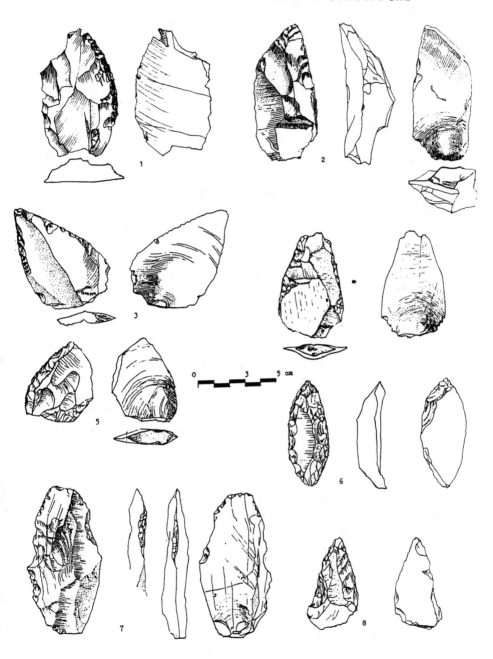

Our techno-typological study (Fistani, 1989) noted some characteristic scrapers (déjeté), limaces and some single choppers; the flakes were of Levallois-technique reflecting a Charentian influence. For this poor sequence we proposed a Riss-Würm chronology. The presence of these Mousterian tools suggests that Middle Paleolithic people occupied North Albania (Fig. 13).

The Mousterian age of these lithic implements was supported by the subsequent study of Gajtan I sediments. From the uppermost layers we recovered an artifactual sequence, though few in number, consisting chiefly of flake tools, but of the same form as those discovered in the valley of Gajtan. Thus it seems that both series are of the same traditional Mousterian Culture. The kinds of lithic elements include scrapers (convexed, right and irregular) and some rare choppers. A dozen elements were discovered in layers C and C_1.

The Middle Palaeolithic is also represented by lithic remains recovered on the surface at the village of Xara (Saranda), near the border of Greece. Flakes were collected and reported by Mustilli (1937-40), and later Korkuti (1982) reported similar findings. At present there has not been a complete study of all of the lithic finds of this open-air Palaeolithic site. Furthermore there are some archaic elements such as choppers in the inventories of this industry that have never been discussed.

Actually there are only some general conclusions that can be drawn about this very interesting open-air site. It seems that the Levallois technique characterizes this industry. The implements are prepared from good raw material. Until now there have been no comparative studies of the North Albanian Mousterian lithic assemblage and those from Xara.

Recently (yet unpublished Fistani, 1986), we recovered a typical artifactual lithic sequence of the Palaeolithic, on the sandy shore of Lake Butrinti, attesting to the presence of two possible human occupations of this clement area during the Middle Palaeolithic. Although the material is still under study, a provisionary report can be made here.

The peoples who occupied this area used pebbles flaked into choppers. We collected two kinds of implements during our study at Butrinti: micro-choppers with a form like the so-called microlithics of the Adriatic, and big choppers produced in nearly the same technological manner. At the same time many flakes were found in the same area. Almost all flakes were made by the Levallois technology.

We suppose that these peoples lived here or in Xara during the last interglacial or Riss-Würm period. Based on known data it seems reasonable to compare this sequence with those of *Strombus bubonius*, discovered in limitrophes in the vicinity.

An essential contribution about lithic industries discovered from the area nearest to South Albania was made by Sordinas, (1973), who carried out research in the islets just offshore near Corfu and Albania (Butrinti). In fact, these islets now belong politically to Greece, but during the Quaternary period the land was connected to that of Butrinti and Saranda. Even now these islets are separated only by shallow water and are of the same geological origin, (Fig. 14). Based on their characteristics, we believe that these widespread industries beginning from Xara, Butrinti and on the isles correspond to the Riss-Würm interglacial, and are associated with *Strombus bubonius*. Typologically speaking, these industries must be considered to represent the same lineage of Middle Palaeolithic people of the Stone Age who were able to migrate throughout this clement area during times of low water levels created by the glacial periods and vice-versa.

The lithic sequences recovered in South Albania near Butrinti display a lack of flake-blades. Up to this time period, such typological elements were extremely rare. The main characteristic is the use of river pebbles transformed principally into choppers or chopping tools. The other elements are flakes prepared by the Levallois technique. Some authors believe that these two typological elements should be divided as separate representatives of two different occupations or two individual cultures, but we suggest that these archaic chopping-tools or chopper elements are a part of the same cultural tradition that also includes the flakes.

A synthesis of all these arguments suggests that in Albania, although the map of Palaeolithic sites contains many hiatuses of unexplored and unknown zones, especially in the central and eastern areas along the peripheral border of the whole country, recent discoveries give cause to believe that there are many sites yet to be discovered. At the same time there is no doubt that a systematic study of the Xara and Butrinti sites — and others as yet undiscovered — would enrich our evidence of Palaeolithic occupation and increase the theoretical probability of discovering human remains from this period.

THE UPPER PALEOLITHIC AGE (FIG. 15)

The ancient Upper Paleolithic Age, or the Aurignacian in a large sense, is represented by a Cultural complex found south-east of Shkoder close to the river Kiri, only 0.7 km from the city. This lithic material was located on the surface of several areas sheltered by a series of hills stretching from west to east on the south bank of the river Kiri. This material is currently under study but its discovery suggests the presence of large groups or populations of hunters and fishermen who sequentially occupied the same areas.

A rich abundance of material (lithic sequences) was discovered on the surface in Bleran, a typical open-air site. A provisionary report suggests to us the presence of an archaic Aurignacian Culture associated with progressive elements, most probably belonging to another cultural sequence. These sequences, in fact, contain qualitatively different raw material — calcareous to high quality silex. Many archaic implements are reminiscent of the ancient stone age and survived to this period as a cultural tradition. They include choppers and chopping-tools which, although rare, were present. Laminas with shaved backs were also present but rare; retouched lamina were not abundant. Lamina Dufour are also found, made from high quality silex. Some typical frontal scrapers are present and one implement resembles both a point and a large retouched back (shaved back). Scrapers with naturally backed flakes are small-sized but typically they are associated with bladelets with fine continuous retouches made on the ventral surface, resembling a kind of lamina Dufour. The implements mentioned above are similar to a type of point discovered in the Polish cave Nietoperozowa of the Upper Palaeolithic of Eastern Europe (see Brézillon and Chmielewski, 1968).

FIGURE 14
A TENTATIVE MAP OF ALBANIA DURING GLACIALS (RISS OR WÜRM) AND ITS VICINITY AREAS (ISLETS OF OTHONIO, CORFU, ETC.)

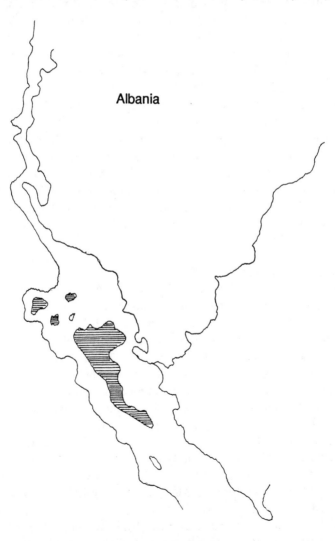

FIGURE 15
BLERAN SITE OF UPPER PALAEOLITHIC - SOME AURIGNACIAN
IMPLEMENTS DISCOVERED IN SURFACE

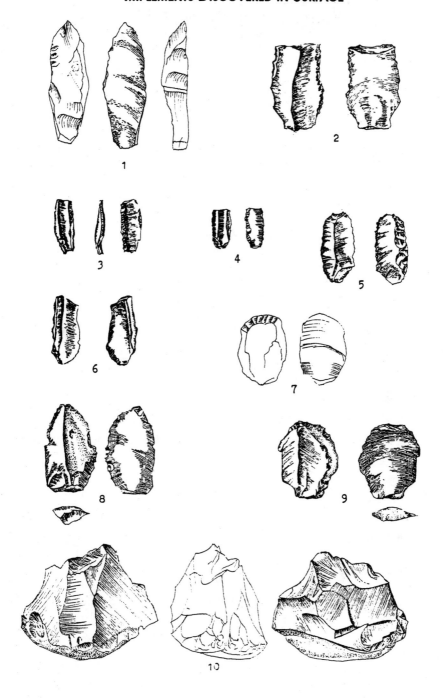

FIGURE 16
MAP SHOWING THE SPREAD OF PALAEOLITHIC SITES FROM ALBANIA
AND THOSE DISCOVERED IN THE VICINITY AREAS IN GITOLJE
(YUGOSLAVIA) AND CORFU (GREECE)
● - CAVE PREHISTORIC SITE
■ - PREHISTORIC SITE IN OPEN-AIR

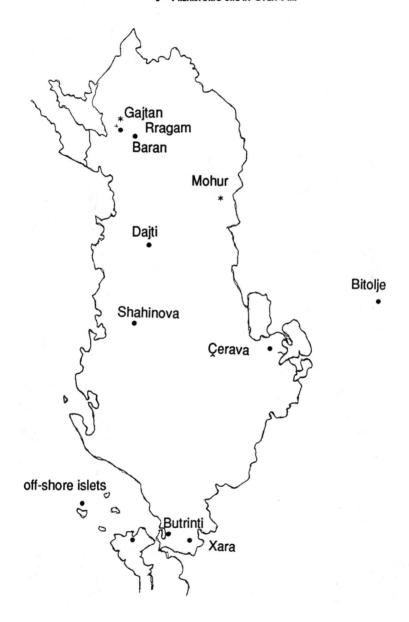

Because the climatic conditions were severe during Würm III (Aurignacian epoch), all settlements were not, strictly speaking, in open-air. There are many reasons to believe that human groups of this period used huts and pitched their tents in the sheltered alcoves of the successive hills of Rrenci. The settlements are all situated facing toward the West, protected from the constant wind and the sharp cold of the Kiri valley. No fossil remains have been found yet that could provide us with information on human morphological adaptation to this paleoclimate.

Evidence of the Aurignacian culture was discovered in the Gajtan II cave, but its presence is more abundant in these open-air sites of Bleran, where successive occupations of Upper Palaeolithic age are evident. Both biological and cultural evolution occurred in these people for survival in the harsh environment in which they lived.

Typologically speaking, the presence of some rare but very thin lamina Dufour, and some typical scrapers or points with shaved back elements show a developed microlithic technique. At the same time the presence of archaic implements such as choppers and heavy duty tools suggest the conservation of a tradition from those earlier cultural groups which evolved in the same areas.

Although the material is under study and our report here is provisional, some similarities can be pointed out with the Catalonian Aurignacian sites which were studied by Masferrer (1977 and 1982). One such case is the Can Crispin Aurignacian site which deserves attention. On the other hand the detection of evolutionary changes within the Albanian Upper Palaeolithic is not so easy to find, because the typological aspects of these implements seem to be closer to the Mousterian tradition expressing a progressive tendency at the same time.

We have demonstrated an Upper Palaeolithic presence in Albania underlining the simple fact that hominids at various stages of human evolution were present in our country. There is evidence that early humans were present during all the geological periods of the Pleistocene. Both cultural and physical evolution of different human groups must have occurred in this area.

CONCLUSIONS (FIG. 16)

The presence of Palaeolithic industries in Albania shows that there are typological sequences of tool traditions representing different stages of the Palaeolithic age. The variety of tool traditions is evidenced by the use of a large variety of raw materials, the ability to adopt a technique for flaking different kinds of hard stones, as is presented in Baran and Gajtan, and the use of the Levallois technique in a gradually evolving manner from Lower to Middle Palaeolithic implements (Gajtan valley) (Gajtan C, C_1). At the same time in Xara and Butrinti, the preparation of both kinds of implements as microlithics — in little pebbles and the preparation of large sized elements — are all cultural traits that evolved during the Middle Palaeolithic or

Lower Palaeolithic.

At present Gajtan I site has not produced fossil hominids. Only the remains of the primate *Macaca*, have been recovered. Still the site is of particular interest, because many of the fossils and artifacts discovered at Gajtan document the first finds of their kind in Albania. Furthermore, this material provides us with much information about human behavioural patterns. Chronologically, the hunting groups belong tentatively to a type of *Homo erectus* which left many traces not only in Gajtan itself, but also in the near vicinity of this prehistoric centre (Baran, Gajtan and Rragam).

On the other hand, the discovery and study of Gajtan I is important because the fossil and artifactual material attests to the presence of prehistoric Man contemporaneous with fossils such as *Macaca, Canis lupus mosbachensis, Dicerorhinus cf. merckii, Cervus elephus, Capreolus capreolus* etc. All these samples represent ancient characters (Fig. 17), for example *Capreolus capreolus* is characterized by a long tooth row and a large basal circumference in the antlers. The presence of *Ursus thibetanus, Dama dama* and *Tortoise* are all characteristic of the warm and moist weather of the Holstenian inter-glacial. Moreover the rich faunal remains are associated with a modest but distinctive archaic artifactual sequence which resembles the industries of the *Trés Ancien Paléolithique* as defined by Bonifay (1989).

FIGURE 17
GAJTAN I (SHKODËR) - SOME FAUNAL ELEMENTS DISCOVERED IN ACHEULIAN LAYERS OF THE SITE

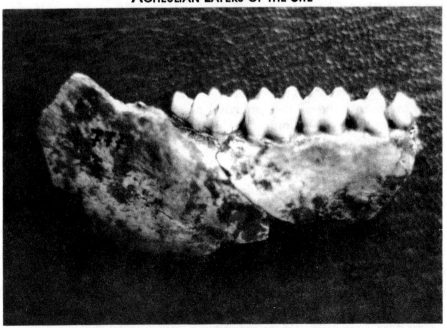

(A) - *MACACA SYLVANA PLIOCENA* — HEMIMANDIBLE NR. 777/GI

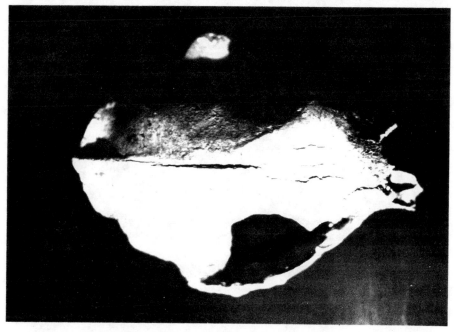

**(B) - Ursus Thibetanus — Discovered in the Cave of Gajtan
Nearly Complete Skull. nr. 3**

**(C) - Ursus cf. Deningeri Discovered in the Cave of Gajtan
Right Maxillary nr. 181**

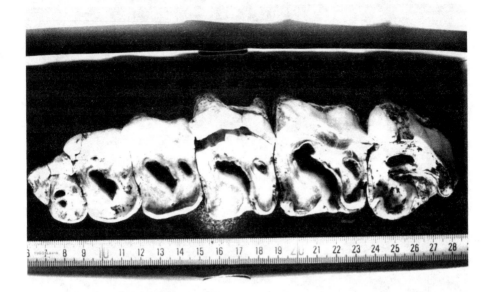

(D) - DICERORHINUS CF. MERCKII FROM LOWERMOST LAYERS OF GAJTAN I
LEFT UPPER TOOTH RAW

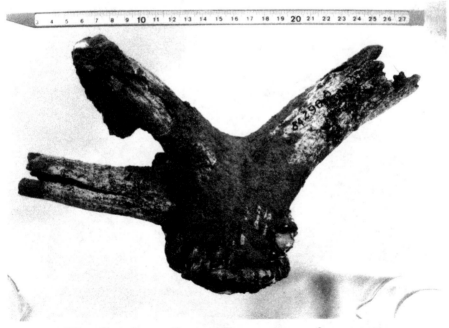

(E) - CERVUS CF. ELEPHUS DISCOVERED IN LOWERMOST
LEVEL OF GAJTAN I — SHED ANTLER, BASAL PART FRAGMENTED
INTENTIONALLY BY PREHISTORIC MAN

It seems possible to assume that the occupation at Gajtan I corresponds to the stage of human evolution that occurred in this part of the Balkan peninsula during the Middle Pleistocene. There are many reasons to give a key position to the Gajtan site from among the other sites discovered in Albania. Although the artifactual spectrum furnished from Gajtan I is not rich, this site has yielded a rich faunal material associated with pebble, choppers, pics, etc. In addition the use of bones and antlers enriches the palaeontological profile of this prehistoric culture.

Because of the richness of artifactual material discovered in the Baran open-air site, compared to other Palaeolithic sites, the typological spectrum is large, certainly in comparison to Gajtan — and enables us to draw some techno-typological observations. Of the total 495 flakes recovered from Baran, 60.2% are flaked from heels covered with primitive cortex, and only 13.9% show prepared heels. The remaining pieces are flaked from their cores. Hence there is poor use of the Levallois technique, and it lies within the inferior limits of an Acheulian industry. There is a predominence of large flakes in this sample. Both crude and archaic artifactual elements are present, and only in rare cases have we encountered well prepared flakes. The index of Levallois technique is IL=25.25 and the index of facets IF-8.48, which again lies near the lower limits for this archaic industry.

It is difficult to find a relationship between Gajtan and Baran, although the two sites are in close geographical proximity. A cultural hiatus of many stages separates these two Palaeolithic sites — the first a semi-rock shelter and the second an open-air site. Secondly Gajtan I yielded poor lithic material with a restricted spectrum of lithic implements; thus the possibility of a parallel comparison between them is difficult. Only the preparation of the bifacial elements are comparable at the two sites. Many typological differences encountered between the artifacts are caused perhaps by the use of different kinds of raw material. In fact every site described here contained different sorts of stone material. In Gajtan I there were pebbles and quartzite, whereas in Baran jasper and calcedonite were used exclusively. Hopefully, future studies can resolve the problem of the cultural hiatus that stands between the sites.

Nearly all of the types of prehistoric cultures mentioned above have been recovered or reported from our region in the North or in South Albania.[6] However due to the lack of a comprehensive and accurate study of all of these industries (most of material is under study), we prefer for the moment to state here that it would be premature to speculate about the relationships of the presently known Albanian prehistoric cultures.

However, there are many promising and intriguing clues to follow up (Fistani, 1989 not published yet). The French author Bourcart (1920), reported the presence of *Hipparion gracilis* and *Equus stenonis* at Cerave (Pogradec) and many of my own cursory inquiries lead me to suspect this area has many secrets to reveal about the *Trés Ancien Paléolithique*.

Archaic pebbles and other artifacts lying in a new open-air site at some of the enigmatic problems of human evolution in this forgotten territory from prehistory. Although not a large area, Albania is very important for the evidence it contains, for it is in this almost abandoned place that oriental and occidental migratory waves of fauna and hominid cultures encountered each other during the Pleistocene.

Future studies will be obliged to work out the problems for each of these cultural epochs in order to understand better how human evolution can be demonstrated in Albania, as in other areas, in this small remaining part of the ancient Illyrian kingdom which in the time of Queen Teuta and Gentius occupied twice as much territory in the Balkan peninsula.

*   ACKNOWLEDGMENTS

The author of this paper wishes to express his sincere gratitude to Dr. Becky Sigmon at the University of Toronto who gave us the opportunity to enter into dialogue and to present for the first time our work of many years and to support it. Furthermore I wish to thank Ms. Isobel Pegg, for her very precious aid in the initial correction of the manuscript.

ENDNOTES

1. All research has been conducted by the author of this paper with the assistance of students in the Biology-Chemistry faculty of University "Luigj Gurakuqi" through 12 years of constant work.

2. M. Naço, A. Papa, M. Korkuti, M. Kabo, A. Dhima, S. Meçe, P. Pashko, Z. Bajrami, J. Therecka, C. Prendi, Gj. Gruda, (23.IV.1982) Protokoll i diskutimeve dhe i shkembimit te mendimeve mbi gjetjet e shpelles se Gajtanit, Shkoder. This material is stored at Laboratori Paleontologjise Humane Shkoder and in the Natural Science Divison of Academy of Science of Albania.

3. The first serious systematic studies in Albania about Ethnic Anthropology were carried out for many years by Cathedra of Normal Anatomy at the Medical Faculty (Ylli, 1975; B. Cipi, 1984). The measurements of living people were compared with skeletal remains recovered during archaeological excavations of QKA of Academy of Science in order to bring to light the Illyrian origin of Albanian people. Dhima, (1982) worked with success for many years in Ethnic Anthropology in the same topic.

4. Following the program of surveys in the vicinity of Gajtan village, we discovered this site; systematic research has been carried out for many years at this site.

5. Some of these palaeonthological elements must be considered as Villafranchian origin, as is the case of *Macaca*.

6. Most of these discoveries are presented in this Symposium for the first time due to the opportunity reserved very kindly by Dr. Becky Sigmon.

REFERENCES

Almagia, Rob., 1918. Trace glaciali nelle montagne dell'Albania. *Revista Georgr. Ital. Anno SSV*, fasc., III, IV, V, Firenze.

———. 1918. Primo Contributo di osservazioni morfologiche sull'Albania centrale. *Soc. Ital. peri il progresso delle scienze*, Roma 1914.

Adam, K.D., 1954. Die Mittelpleistozänen Faunen von Steinheim an der Mur (Württemberg), *Quaternaria*, I. 131-144.

Argant, A., 1989. Carnivores quaternaire de Bourgogne. Thèse de doctorat. Université Cl. Bernard, Lyon I, no. 152—89, Nr. 5 (*Ursus thibetanus*).

Baldacci, A., 1902-1904. Nel Paese del Cem. (Viaggi di esplorazzione del Montenegro orientale e sulle Alpi Albanesi), *Bolletino di Societa Geografica Italiana*.

Basler, D., 1967. Arheoloski nalazi u Crvenoj Stijeni, Glasnik. *Zem. Muz.* n.s. XXI/XXII, Sarejevo.

Behrensmeyer, A.K., et al., 1986. Trampling as a cause of bone surface damage and pseudo-cutmarks, *Nature*, vol. 319-27 (February.

Berciu, D., 1945. Paleolitiku dhe Mezolitiku ne Ballkan (Leksion i mbajtur ne Fakultetin e Filizofise dhe te letrave ne Bukuresht, 1945). *Inst. Arkeologjik Rumun ne Shqiperi, Illyrica*, I.

Bonifay, E., 1989. Les prémieres industries préhistorique en Europe, in Le Temps de la Préhistoire, tome 1, *Assoc. Préhist. Franç*. Edition Archaeologia.

Bonifay, M.F., 1971.Carnivores quaternaires du Sud-Est de la France, *Mém. Mus. nat. Paris*, n.s. série C, T. 21., fasc, 2, p. 43-377.

Bordés, F., 1950. Principe d'une methode d'étude des techniques de débitage et de la typologie du paléolithique ancien et moyen., *L'Anthropologie*, t. 54.

———. 1988. "Typologie du Paléolithique ancién et moyen" *presse du C.N.R.S.*, 10-101, PL. 108.

Boué Ami, 1840. *La Turquie d'Europe*, Paris.

Bourcart, J., 1920. La glaciation quaternaire dans l'Albanie moyenne *Compte rend.*, d.l'Ak.d. sc. Paris.

———. 1916-1920. *Les confins Albanais administrés par la France*, Paris, 1922.

———. 1925. Sur la stratigraphie de l'Albanie orientale, a propos des deux notes de M.E. Nowack, *Bull. Soc. Géol. d. Fr.* 5. 25, Paris.

Brace, C.L., 1964. The fate of the classic Neanderthal's: a consideration of hominid catastrophism, *Current Anthropology*, 5:3-43.

Brézillon, M., (1977, La dénomination des objets de pierre taillés, IV, supplement a *Gallia Préhistoire, Edition du C.N.R.S.* Paris, pg. 1-423.

———. and Chmielewski, 1968. *Civilisation de Jerzmanaowice, Zalclad Narodowy Imenia Ossolinskich Wydawnitotwo Polskiej Akademii Nauk.*

Caloi, L., Palombo, M.R., 1978. Anfibi, rettili e manniferi di Torre dal Pagliaccetto (Torre in Pietra, Roma), *Quaternaria,* XX. a cura di A. Malatesta, 388-400.

Canal, J., Masferrer, N.S., 1976. *El Paleolitic a la Comarques Gironinen, Caixa d'Estalvin de Girona.*

Cipi, B., 1984. Probleme Anthropologjike, (Perdorimet e tyre osteologjike ne pergjithesi dhe ne praktiken mjeko-ligjore e kriminalistike ne vendin tone), UT, Tirane, Fak. Njeks. 1-227, (dissertation).

Chavaillon, J. et alii., 1978. Le debut et la fin de l'Acheuléen à Melka Konturé: méthodologie pour l'étude des changements de civilisations, *Bull. Soc. Préhist.* Fr. 75,4.

Crégut-Bonnoure, et S. Gagniere, 1989. Le problème de l'existence d'élements extreme-orientaux dans la faune du Pléistocène Européen: découverte d'Ursus thibetanus (Mammalia, Carnivora, Ursidae), dans le site de la Baume Longue (Dions Gard, France). *Revue de Paléontologie,* vol. 8, nr. 1, ISSN, 0253-6730, 65-71, Génève.

Cvijiç, Jov., (1917. L'époque glaciaire dans la peninsule balkanique, *Ann. de Géogr.,* XXVI, 141-142.

———. 1901. Die dinarich-albanesische acharung. d. kais. *Akad. d. Wiss.* CX, Wien.

Day, M.H., 1986. *Guide to Fossil Man,* 4th edition, the University of Chicago Press, Chicago, pp. 1-465, fig. 133.

Degrand, A., 1901. "Souvenir de la Haute-Alvanie" 256-258.

Delson, E., 1980. Fossil Macaque, phyletic relationships and scenario of deployment. *The Macaque: Studies in Ecology and Behaviour* ed. D.G. Lindburg; 1980, Van Nostrand New York, 10-30.

Dhima, A., 1982. Te dhena antropologjike per gjenezen e shqiptarve, (dissertation).

———. 1977. Disa rezultate paraprake ne studimin e etnoantropologjik te popullit tone. Konf. Komb. e stud. Etnografike, Akademia e Shkencave, Tirane, fq. 543.

———., 1979-1980. A propos du type anthropologique des albanais durant le Moyen Age, 301-142, *Iliria* IX-X, Tirane.

Dodona, E., Kotsakis, T., 1985. Prémiere découverte de Mastodontes (Proboscidae, Mammalia) en Alganie. *Geologia Romana,* nr. 24, (1985): 73-78, vol. XXIV, 73-78, Roma.

Erdbrink, B.D.P., 1981. The oldest stone tool of N.W. Europe, serie B, 84, (2), University of Utrecht, Netherlands.

———. 1983.Sundry fossil bones of terrestrial mammals from the bottom of the North Sea., *Palaeontology, Proceedings,* B., 86 (4), December 19.

Fistani, A., 1986. Gjurmë të aktivitetit parahistorik në vëndgjetjen shpellore të Gajtanit (Shkodër), *Buletini Shkencor* nr. 2, Instituti i Larte

Pedagogjik, 83-94.

_____. 1987.*Canis lupus mosbachensis* (Soergel) fosil nga shpella e Gajtanit (Shkodër), *Buletini i Shkencave te Natyres* nr. 1, 115-122., 6 tab. Tirane.

_____. 1987. Gjetje fosilesh me interes paleoetnologjik nga shpella e Gajtanit (Shkoder), *Buletini Shkencor*, nr. 1:117-131. Shkoder.

_____. 1990.*Ursus Deninger von Reicheneau* nga shpella e fosileve Gajtan (Shkoder), *Buletini i Shkencave Natyrore*, 4/1990, 53-63, fig. 4, pasq. 3.

Glüeck, L., 1897. Zur Physischen Anthropologie der Albaneser, *WMBH*, 364-366 and 376-402.

Gabori, M., 1976. *Les Civilisations du Paléolithique Moyen entre les Alpes et l'Oural*, Budapest.

Girard, C., 1978. Les industries Mousteriennes de la Grotte de l'hyene à Arcy-sur-Cure (Yonne), XI supplement à *Gallia Préhistoire, edition du C.N.R.S.*

Gobineau, A. comte de., 1853. *Essais sur l'inégalité des Races Humaines.* II édition, Firmin-Didot, Paris.

Haberland, A., Lebzelter, V., 1919. Zur Physischen Anthropologie der Albaner, in *Archiv für Anthropologie*, 123, XVII.

Hamy, M.E.T., 1900. Contribution à l'Anthropologie de la Haute-Albanie. *Bulletin du Musée Hist. Nat.* Paris, 169-172.

Hill, A., 1986. Tools, teeth and trampling, *Nature*, vol. 319-, 27 February.

Koby, F. Ed., 1938. Une nouvelle station préhistorique (paléolithique, néolithique, âge du bronze): les cavernes de St-Brais (Jura Dernois), pg. 138-194, Verhandlungen der Natur-schenden Gesellschaft in Basel, band XLIX, 1937-38.

Kurtén, B., 1959. *On the bears of the Holsteinian Interglacial*, Stockholm Contr. Geol. 2, Sweden, 75.

_____. 1980. Fossil Carnivora of Petralona Cave, status of 1980, *Anthropos*, tom. 80.

Korkuti, M., 1982. Gjetje të reja të paleolitit të mesëm nga stacioni i Xares (Sarande), *Iliria*, nr. 1, 39-49.

Lebzelter, V., 1919. Ein albanischer Schädel aus der Völkerwander ungszeit *Archiv für Anthropologie* XVII, Braneschweig 1919, 3-4 142-146.

Lumley, H. de, 1971. Le Paléolithique inférieur et moyen du midi méditerranéen dans son cadre géologique, tome I et II, Ve supplément à *Gallia Préhistoire, edit. C.N.R.S.*, Paris.

Miracle, P., and Sturdy, D., 1991. Chamois and the Karst of herzegovina. *Journal of Archaeological Science*, 1991, 118, 89-108.

Milaj, J., 1944. "Raca Shqiptare", (studim antropologjik historik), Ismail Mal Qemali, Botonjës — Tirane, 1-151.

Mustilli, D., 1937-1940 *Relazione Preliminare Sugli Scavi Archeologicici in Albania*, RAI, *Rendiconti delle classe di scienze morali e storiche*, II, Roma.

_____. 1940. *La civilta Prehistorica dell'Albania*, estratto della revista d'Albania, vol. XVIII, Milano.

————. 1954-55. Ricerche italiane per la preistoria dell'Albania Bolletino di *Paletnologia italiana*, II, vol. 64, Roma, 401-409.

————. 1965. Të dhëna arkeologjike për prehistorihë e Shqipërise Konferenca e pare e studimeve Albanologjike, (15-21 Nëntor 1962, Tirane), 457-459.

Nopsca, Fr., 1911. Sind die heutigen Albanesen die nachkommen der alten Ilyren, *Zeitsch.*, *Ethn.*, Berlin, n. 6.

————. 1925. Zur Geologie der Kustenketten Nordalbanies., *Mitt. D. Jalrb. d. kgl. und., Geol.* Anst., 24 Budapest.

Nowack, E., Dr., 1929. Eiszeitliche Bildungen in Albanien, *Zeitschr. f. Gletscherkunde.*

Obermeier, H., 1928. Dereselbe, Das Paläolithikum des Balkan. *Eiszeit u. Urgeschichte*, 5, 1928, s. 24.

Patte, Et., 1918. Coup de poing, en quartzite, des environs de Monastir (Serbie), *Bull. Soc. Préhist. Franç.*, 15, (1918).

Puttard, E., 1920. *Les Peuples des Balkans*, Paris, 81: 298.

Piveteau, J., 1957. Traité de Paléontologie, tome VII, Primates, *Paléont Humaine*, fig. 639, 1-670.

————. 1921. *Découverte de l'âge de la Pierre en Albanie.*

Ray Leon, 1928. *Albanie, Cahier d'Archaeologie d'Histoire et d'Art en Albanie et dans les Balkans*, MCMXXVIII, nr. 3 — Repertoire topo-bibliographique des Antiquité de l'Albanie.

Reshetov, Ju. G., 1966. *Priroda Zemli i Proishozhdenie Çelloveka*, 326 (Centralnoevropejskoj-Ballkanskaja obllast). Moskva.

Richtoffen, B. von., 1939. Die ersten Spuren der Eiszeitmenschen in Albanien, *Quartar*, vol. 2, 121-152, Bonn.

Roquet and Maferrer, 1976. *El Paleolitic a la Comarques Gironines*, CAIXA d'Estalvin de Girona. Girona Espagne.

Sigmon, B.A., 1989. Locomotor Adaptations in *Homo erectus*, *EAZ Ethnogr Archaöl, Z.*, 30, 501-514.

Skutil, J. Vig., 1950. L'époque quaternaire dans les Balkans, Compte Rendu du III, Congrès des Géopgraphes et Ethnographe Slaves en Yugoslavie, Belgrade.

Soler i Masferrer, N. 1977. El jaciment preistoric de Can Crispin, 7-38, fig. 26, *Cypsela* vol. II, Girona.

————. 1982. El jaciment prehistoric de Can Crispins i l'Aurinyacia de Catalunya, 7-12, fig. 18, *Cypsela* VI, (Llagostera Girona).

Sordinas, A., 1973. Stone Age Sites on Off-Shore islets Northwest of Corfu, Greece, 72nd Annual Meeting of the American Anthropological Association New Orleans, Lousiana U.S.A.

Tillier, A.M. et Vandermersche, B., 1976. Les cynomorphes (in Lumley H. de. La Préhistoire Française), Paris, C.N.R.S. 367-370. fig. 2.

Ugolini, L.M., 1927-1942. *Albania Antica*, in 3 vol.

Virchow, R., 1877. Zur Kraniologie Illyriens Monatsbericht der Königlichen Akademie der Wissenschaften zu Berlin, 768-803.

Weininger, J., 1934. Rassenkundliche Untersuchungen an Albanern; ein Beitrag zum Problem der dinarischen Rasse. In Rudolf Pöchs Nachlass, Serie A; Physische Anthropologie, Bd. IV, *Anthropologische Gesellschaft*, Wien, vol. 4, 1-68.

White-Montet, A., Laville, H., Lezine, A.M., 1986. Le Paléolithique en Bosnie du Nord, chronologie, environement et préhistoire. *Anthropologie*, (Paris), tom. 90, nr. 1, 29-88.

Ylli, A., 1975. *Disa te dhena kraniometrike*, 1-151, Tirane.

Zampa, R., 1886. Anthropologie illyrienne *Revue d'Anthropologie*, IX (2e serie) Paris, 625-648.

CHAPTER 10

PHYSICAL ANTHROPOLOGY IN POLAND: PAST, PRESENT AND FUTURE DIRECTIONS

J. PIONTEK

INTRODUCTION

I. SHORT HISTORY OF PHYSICAL ANTHROPOLOGY IN POLAND

The history of the development of anthropology in Poland has been well described. In 1956, a hundred years had elapsed since the introduction of the first (both in name and content) lectures in anthropology in the world at the Jagiellonian University in Cracow. These lectures were given by J. Majer who created the Anthropological Commission at the Academy of Science in Cracow in 1873.

The hundred-year's anniversary of university teaching of anthropology was a good opportunity for writing numerous monographs discussing the output of various universities.

Two works discussing the history of anthropology in Poland appeared in 1985: Piontek and Malinowski's book, "Theory and Empiricism in the Polish Anthropological School," and Malinowski and Wolański's "Anthropology." The latter article on the history of Anthropology in Poland discusses the present state of anthropology in Poland, directions and the more important results of research, anthropomorphology of skeletal materials and living populations, human growth and development, human ecology and other divisions of physical anthropology.

This article presents the history of the development of particular fields of physical anthropology in Poland. It also discusses the most important results of studies conducted until 1980. It is intended to be a very thorough work containing:

1. information about all the centers of anthropological studies;
2. information about the development of anthropological institutions;
3. information on the authors of studies and the main directions of anthropological studies in Poland in particular periods of development of anthropology.

The authors did not give the bibliography of the major works by Polish anthropologists because Wrzosek (1956) and later Godycki (1967; 1970; 1976; 1981) compiled a whole list of works by Polish anthropologists. A bibliography of Polish works on anthropology written until 1956 was published by Wrzosek (1956) as a separate volume of the journal *Materialy i Prace Antropologiczne* (vol. 41). A bibliography for the years 1956-1980 was prepared by Godycki as separate volumes of the journal *Przegląd Antropologiczny* (vol. 55, 1992).

Another work entitled *History of Physical Anthropology in Poland* was written by Bielicki, Krupínski and Strzalko (1985). It was published by the International Association of Human Biologists. This work discusses the development of anthropology in Poland from the historical point of view.

The work discussed here is more than a chronicle of the history of development of anthropology in Poland. Apart from historical data it also contains: (a) a list of the main trends in anthropological studies, (b) critical reflections, and (c) short remarks on the historical and political conditions of the development of anthropology in Poland. The author devoted a lot of space to: (a) the history of the creation of The Polish Anthropological School connected with the activity of Czekanowski (1962), (b) criticism of the "typological" direction and (c) historical conditions of the birth of the modern (post-typological or better populational) anthropology in Poland.

Due to the manner of interpretation of problems, good selection of the main information and full presentation of the main directions in anthropological studies, it is hoped that this article will be a useful and interesting summary of the history of anthropology in Poland.

II. THE PRESENT STATE OF ANTHROPOLOGY IN POLAND: INSTITUTIONS[1]

In 1987, T. Krupínski presented the results of a poll concerning the following information:

1. number of people active in anthropology;
2. scientific position;
3. place of work;
4. scientific interests.

It can be seen from the 1987 survey that in Poland 183 anthropologists were active, among them 9 professors emeriti. In total, there were 42 professors, 61 adjunct professors, 6 lecturers, 39 assistants and 23 technical assistants. Anthropologists work in the following institutions: universities, higher pedagogical schools, physical education academies, medical academies, the Polish Academy of Science and archaeological institutions. The largest centres of anthropology are in Wroclaw (28% of all anthropologists are there), Warsaw (23%), Poznań (18%), and Cracow (9%).

The most interesting subjects are those problems which concern human ontogenesis, human morphology, skeletal biology, genetics and applications of anthropology in sport and physical education. It is symptomatic that those disciplines which are well-developed in modern anthropology such as anthropogenesis, human ecology, primatology, or sociobiology do not incite more interest. Ethnic, social and cultural anthropology are also marginal in the interests of Polish anthropologists. The scope of scientific interests of Polish anthropologists is not reflected in the works published by Polish anthropologists in international journals. The largest number of works published in English during the last five years deals mainly with social conditions of human physical development, genetics of quantitative traits, cultural anthropology, biology of prehistorical populations and human ecology.

The best developed centres, both from the point of view of scientific level and organization, are the Institute of Anthropology of the Polish Academy of Science in Wroclaw, the Institute of Anthropology of the University in Poznań, the Department of Historical Anthropology of the Warsaw University, the Department of Anthropology of the Jagiellonian University in Cracow and Wroclaw and the Departments of Anthropology at the Academies of Physical Education in Cracow, Poznań, Warsaw.

In Poland scientific institutions are divided into:

1. a special governmental institution;
2. universities and higher schools conducting teaching and research activities in co-operation with the PAN and other institutions;
3. so-called "departmental" or "ministerial" institutes, conducting mainly applied research projects for various branches of economy, industry, health care, environmental protection, education, culture etc.

Anthropology classes are held at universities for students of biology, ethnology, archaeology, psychology, pedagogy, and at Physical Education Colleges and at Medical Colleges within a framework of anatomy and pediatrics.

Specialists in anthropology are trained at universities in Cracow, and Poznań Wroclaw. Doctoral and post-doctoral studies are available in the Institute of Anthropology PAN in Wroclaw and the Institute of Anthropology in Poznań , and at Departments of Anthropology in Cracow and Wroclaw.

Departments of Anthropology affiliated with Faculties of Biology are established at Universities in Lódź (Head: A. Malinowski), Cracow (Head: K. Kaczanowski), Poznań (in Poznań 6 anthropological departments operate in the Institute of Anthropology: Department of Biology of Human Development - Head: J. Cieślik; Department of Human Populational Ecology - Head: J. Strzalko; Department of Applied Anthropology and Ergonomics - Head: A. Malinowski; Department of Human Evolutionary Biology - Head: J. Piontek; Museum and Osteological Laboratory - Head: E. Milosz), Torun (Head: G. Kriesel), Wroclaw (Head: T. Krupinski), Szczecin (Head: Z. Szczotka and Warsaw (Head: A. Wiercinski).

Departments of Anthropology are also established in the Physical Education Colleges in Warsaw (J. Charzewski), Cracow (Head: S. Golab), Wroclaw (Head: A Janusz), Poznań (Head: Z. Drozdowski), Gdansk (Head: J. Gladykowska-Rzeczycka); and the Anthropological Laboratory in Katowice and in Gorzow Wlkp.

Anthropological laboratories are active in Medical Colleges at the Departments of Anatomy in Bialystok, Katowice, Poznań , Wroclaw, Szczecin. Some problems of anthropology also constitute the main part of the activity of the Department of Anatomy in Szczecin, Bialystok, Gdansk, Cracow and Lódź. In the Institute of Pediatrics, anthropological problems are studied in Lublin, Cracow, Lódź, Warsaw, and in the Institutes of Orthodontics, in Poznań , Wroclaw and Cracow.

There are also anthropology laboratories present at the Higher School of Education in Warsaw, Czestochowa, Slupsk and Kielce. In these schools, as well as at the Faculty of Pedagogics of universities (in 21 university schools in all) *Biometrical Bases of Development and Education* exists containing much anthropological information.

In the Polish Academy of Science two anthropological institutions operate.

In the Institute of Anthropology of the PAN in Wroclaw (Director: T. Bielicki) some 25 anthropologists are employed, among them, Professors P. Bergman, B. Hulanicka, A. Orczykowska-Swiątkowska, B. Miszkiewicz, Z. Welon, E. Piasecki.

In the Warsaw Department of Human Ecology at the Institute of Ecology PAN about 5 anthropologists are employed (Head: N. Wolański).

INSTITUTE OF ANTHROPOLOGY, UNIVERSITY OF POZNAN (POZNAN, UL. FREDRY 10)

Principal research projects:

1. **Human biology** - theoretical aspects of ontogeny, prenatal development, factors influencing growth and development, elaboration of measures and developmental norms, secular changes

in human growth, influence of hormonal disturbances on developmental processes;

2. Populational biology and human ecology - theory of human evolution, biocultural adaptive mechanisms, quantitative evaluation of intensity of various evolutionary factors operating in human populations, genetics of quantitative characters;

3. Historical anthropology - state and dynamics of prehistoric populations and historical records, biology of local human groups in prehistorical and historical times, description of ways of intra- and intergroup differentiation, paleopathology, paleonutrition, stress in prehistorical populations;

4. Morphology and morphogeny of the human skeleton - development of bones, variability of morphological characters, elaboration of criteria for sex and age estimates on skeletons; and,

5. Applied anthropology and ergonomics - criteria and methods of assessing labour safety and ergonomics for technological processes and in using machines and other equipment in food industry, study of work and of organization of specific workshops and factories.

The Institute of Anthropology at the Faculty of Biology in Poznań University is the largest anthropological research centre in Poland. In it is housed the editorial office of the oldest Polish anthropological journal *Prezeglad Antropologiczny* and of the Journal of the Faculty of Biology *Variability and Evolution*.

Some of the most important results are as follows:

1. Publishing of manuals and readers for students in the fields of archaeology, ethnography, physical anthropology, human biology, human ecology, methods of anthropological research, and general anthropology.

2. Elaboration of the studies on the state of the biological development of populations in north-western Poland and presentation of theoretical and methodological suggestions for studies of the processes of human development.

3. Development of theoretical-methodological foundations of studies in biology and ecology of subfossil populations, and in particular (a) measures of the state and biological dynamics of prehistorical populations, evaluation of the possibilities of influence of natural selection in subfossil populations, (b) working out of the role of technological and organizational transformations (Neolithic and industrial revolution) in the formation of the biological structure of human populations.

4. Attempting to test empirical material of some socio-biological theorems.

5. Presentation of theoretical-methodological suggestions for paleoecological analyses and of an attempt at description of the ecological structure of subfossil populations on the basis of model interpretations and analyses of anthropological and archaeological data.

6. Attempts at applications of physical anthropology for the ergonomics problems and labour ecology.

INSTITUTE OF ANTHROPOLOGY OF POLISH ACADEMY OF SCIENCE (WROCLAW, UL. KUZNICZA 35)

The principal research projects include:

1. social stratification and secular trends;
2. individual variation in growth patterns: longitudinal studies;
3. methods of constructing growth standards;
4. the genetics of growth: longitudinal studies of twins;
5. variability and genetics of anthroposcopic features;
6. engineering anthropometry;
7. biodemography; and,
8. skeletal biology.

Some of the most important results are:

1. Description and explanation of the role of social gradients in the formation of biological variability of the contemporary population of Poland.
2. Elaboration and explanation of the processes of biological development on the example of unique materials concerning twins;
3. Longitudinal studies of the phenomena of human biological development.
4. Designing anthropometric standards of the population of Poland.
5. Monitoring of the biological state of population of Poland.
6. Elaboration of the role of psychic stresses in the process of human development.
7. Biology of skeletal populations from Poland.
8. Theoretical and methodological suggestions for studies of the processes of human growth and development.

DEPARTMENT OF HISTORICAL ANTHROPOLOGY IN INSTITUTE OF ARCHAEOLOGY AT UNIVERSITY OF WARSAW (WARSZAWA, UL. KRAKOWSKIE PRZEDMIESCIE 26/28)

Principal research projects are:

1. anthropological and systemic definition of culture, especially of Man as a species;
2. modelling of human biological evolution: ontogenetic retardations and human evolution, evolutions of human psychic functions, the origin and evolution of articulated speech;
3. ecosensitive variability and microevolutionary variability of *Homo sapiens*;
4. application of historical anthropology for ethnogenesis: the processual and structural analysis of palaeoethnic affinities in historical anthropology, and anthropological contribution to the origin of Slavs;
5. religious systems of ancient Mexico and their origin; and,
6. cultural development in pre-Columbian Peru.

Some of the most important results are:

1. theoretical studies in the field of cultural and general anthropology, including proposal of anthropological definition of notion of culture, anthropological conception of ideological development of mankind, anthropological interpretation of the peculiarity of species nature of man;
2. description and analysis of microevolutionary processes in the Holocene;
3. studies in the field of cultural anthropology concerning the oldest centres of civilization (Egypt, Mexico, Peru); and,
4. biocultural analysis in the evolution of *Homo sapiens*.

DEPARTMENT OF ANTHROPOLOGY, JAGIELLONIAN UNIVERSITY (KRAKOW, UL. KARASIA 6)

Principal research projects are:

1. paleoanthropology;
2. primatology;
3. human cariology; and,
4. biology of human development.

Some of the most important results are:

1. studies of biological development of children in industrialized areas;
2. studies of the biology of populations in Southern Poland;
3. studies of the biology of subfossil populations (morphology, paleodemography, paleopathology);
4. experimental studies in the field of primatology; and,

5.   human variability and taxonomical consequences.

DEPARTMENT OF ANTHROPOLOGY, UNIVERSITY OF LODZ (LODZ, UL. BANACHA 12)

Principal research projects are:

1. biology of human development;
2. historical anthropology;
3. ethnic anthropology.

Some of the most important results are:

1. elaboration of microevolutionary transformations in subfossil populations;
2. studies in the field of biology of development;
3. studies of the morphological structure of various ethnic groups from Asia, Africa and South America.

DEPARTMENT OF ANTHROPOLOGY, UNIVERSITY OF TORUN (TORUN, UL. GAGARINA 9)

Principal research projects are:

1. biology of skeletal populations (morphology, paleodemography, paleopathology, methods of analysis);
2. inheritance and differentiation of dermatoglyphic traits;
3. elaboration of the state of biological development of children from Cuiavia and Pomerania (Poland regions);
4. Methodological research in the field of historical anthropology.

Some of the most important results are:

1. biological differentiation of skeletal populations, state and biological dynamics of the prehistoric human groups;
2. studies in the field of the secular trend;
3. inheritance of dermatoglyphic traits;
4. body build in prehistoric populations.

DEPARTMENT OF ANTHROPOLOGY, UNIVERSITY OF WROCLAW (WROCLAW, UL. KUZNICZA 35)

Principal research projects are:

1. studies of the biological structure of population of southwestern Poland;

2.   the secular trend in the biological development of children;
3.   anthropogenesis;
4.   studies of the physical development of African children; and,
5.   historical anthropology,

Some of the most important results are:

1.   creation of the collection of reconstructions of the fossil Hominidae forms;
2.   studies in the field of the secular trend (inter-generational changes);
3.   studies concerning evaluation of the physical development of children from Africa; and,
4.   elaboration in the field of historical anthropology.

DEPARTMENT OF HUMAN ECOLOGY, ACADEMY OF SCIENCE (WARSZAWA, UL. NOWY SWIAT 72)

Principal research projects are:

1.   research on human populations in various parts of Poland are interpreted as bio-indicators of environmental conditions. Studies have been made in the agricultural rural areas of the north east (Suwalki region), in rural areas in the east where industrialization is just beginning (Lublin coal basin), in small towns in the centre of the country where industrialization has been proceeding for some time, in the heavy industry centres of coal mining and metallurgy in the south (Silesia), and in a large city in the centre of the country where the textile industry is situated (Lódź);
2.   the Department has also carried out biological comparisons of human populations in villages and the City of Warsaw. Similar comparisons have been made for the populations living under different climatic conditions, contrasting the Hel peninsula on the Baltic Sea with the mountains of Pieniny in the south. The records in the department contain data on about 11, 000 two to three generation families from these populations.

The most important results are:

1.   some of the monographs have been reprinted; *The Biological Development of Man* is already into six editions (1970-86); *Factors of Human Development, Introduction to Human Ecology.* Since 1973 the department has also published *Studies in Human Ecology* and volume 10 is currently in preparation; and,
2.   the most important problems are inter-familial and inter-populational differences in relation to socio-economic factors; health

status in rural and urban areas in relation to living conditions, climate and mode of life; factors affecting fertility and survival in different populations; marriage distance, assortative mating and self regulatory mechanisms in human populations; respiratory, cardiovascular and haematological characters in relation to differences in body size and composition in varying environmental conditions.

III. THE PRESENT STATE OF ANTHROPOLOGY IN POLAND: DIRECTIONS OF RESEARCH AND MORE IMPORTANT RESULTS

1. HUMAN EVOLUTION

Craniological and odontological studies of Primates and fossil Man were conducted by W. Stęślicka, F. Rosiński, T. Krupiński, Z. Rajchel.

A new concept of mechanisms of hominization was presented by T. Bielicki in 1969 (*Some Feedback Relations in the Evolution of Hominidae*). Questions of the rate of phylo- and ontogeny of the cranium were dealt with by A. Wierciński.

A special line of paleoanthropological research has for many years been pursued by T. Krupiński and Z. Rajchel: anatomical reconstructions of skulls of fossil hominids and hominoids from incomplete specimens.

One recent development worthy of note was the setting up in 1983 of an informal multidisciplinary study group, organized and chaired by A. Wierciński at Warsaw University and A. Wiercińska at the State Archaeological Museum in Warsaw, grouping together several physical anthropologists, ethnologists, experimental psychologists and computer scientists. The group's task has been to explore, from various angles, the problem of *The Peculiarity of Man as a Species*. It is perhaps symptomatic of certain more general trends and moods which surfaced in Poland in the 1980s that a conspicuous tendency making itself felt during the discussions held by this group was a nostalgic search for "real discontinuities" and " quantitative differences" between Man and the rest of the animal kingdom.

2. SKELETAL BIOLOGY

A topic which during the 1960s and 1970s attracted considerable attention of a number of researchers in Poland was brachycephalization and the problem of microevolutionary changes in cranial morphology in the Holocene. In particular: the hypothesis of a change in cranial morphological structure of human populations after the "neolithic revolution."

Extensive work has been done in the field of paleodemography, paleopathology, ethnogenesis of Slavs and morphology of the prehistoric and historic populations from Central Europe. Considerable attention has also been devoted to methods of reconstructing the demographic characteristic of prehistoric and early historic populations both by refining the techniques of determining sex and age at death (also from cremated materials, and also critically examining the demographic relevance).

Microevolutionary processes taking place during past centuries were objectives of works carried out by A. Wierciński, T. Bielicki and Z. Welon, M. Henneberg, J. Piontek, and others. A synthetic monograph *Microevolutionary Processes in European Human Populations* has been published recently by J. Piontek (1979). Human populations treated as biological systems are studied by M. Henneberg, J. Piontek and J. Strzalko (1980), who analyse evolution of populations with respect to their paleoecology as well. Characteristics of their works were a strong methodological orientation, as exemplified by the search for new measures of opportunity for natural selection (M. Henneberg, J. Piontek 1975), and also a strongly integrative approach. A synthetic monograph *Biocultural Perspectives on Ecology of the Prehistoric Populations from Central Europe* was published by J. Piontek and A. Marciniak (1990), and *Ecology of Prehistoric Population* was published by J. Strzalko and J. Ostoja-Zagorski.

3. Human ecology

This area has long been the domain of a very active team organized by N. Wolański (Department of Human Ecology of the Academy of Science in Warsaw). The works of N. Wolański's group cover different fields of human biology: from human ecology *sensu stricto* (the influence on growth of nutrition and physical factors of the environment), various aspects of population structure (e.g., assortative mating, mating distance, heterosis), the genetics of continuous traits (blood pressure, auricular morphology, physiological traits), clinical standards of growth, urban-rural differences in growth, secular trends in growth, variation and age-changes in body posture, motor development in children and adults, physical work capacity, natural selection in contemporary man — up to and including broad theoretical considerations of problems such as genotype-environment integration, canalization of growth, and secular trend versus micro-evolution.

4. Auxology: biosocial stratification and secular trends

The population of Poland constitutes a particularly interesting object for such studies. For example, post-war Poland, in sharp contrast to the pre-war situation, has been one of the ethnically most homogeneous countries in Europe, with no linguistic, religious or racial minorities of any numerical significance.

This area has long been the domain of a group organized by T. Bielicki (Institute of Anthropology of the Academy of Science in Wroclaw). More important results are: (1) the effects of urbanization on growth were analysed not only in terms of the simple urban-rural dichotomy but also with the aid of much more finely graded scales, ranging from big cities through several levels of smaller cities and towns down to villages, (2) urbanization and parental education have each a significant effect *per se* on statures of both children and conscripts, (3) secular trends in growth and maturation have been intense in post-war Poland, the urban-rural gap has shown little tendency to decline, rural children are inferior in growth to even the children from neighbouring small non-industrial towns, (4) in one big city population parental education has decreased in importance as a "stratifying agent" whereas family size has not.

These and other findings of this sort have been described in many reports published in international journals, the Polish anthropological journal *Studies in Physical Anthropology, Materialy i Prace Antropologiczne* and also in a special series *Monographs of the Institute of Anthropology, Polish Academy of Science.*

5. AUXOLOGY: INDIVIDUAL VARIATION IN GROWTH PATTERNS

Auxological research constitutes one of the oldest areas of interest in Polish anthropology. Developmental studies are conducted in all academic centres. Most of these studies are concerned with the development of somatic characters observed in longitudinal studies and cross-sectionally on school children.

Two major longitudinal studies of non-twins were carried out during the post-war period, one in Warsaw, another in Wroclaw.

Among the many analyses of that material the following must be mentioned:

(1) the demonstration that children who throughout their development consistently differ in Parnell method somatotype follow different developmental pathways with regard to sexual and skeletal maturation and growth in size;

(2) an analysis showing how the relative importance of genetic and environmental influences on growth can be estimated from the pattern of inter-age correlations;

(3) a follow-up study designed to find out whether or not the marked difference in the tempo of maturation during adolescence (early vs. late maturers) has any long-term effects on the educational, occupational and marital histories during early adulthood.

6. AUXOLOGY: LONGITUDINAL STUDIES OF TWINS

Two studies of monozygotic and dizygotic twins, both longitudinal, were launched independently in 1966-1967, one by M. Sklad at the Academy of Physical Education in Warsaw, another by Z. Orczykowska at the Wroclaw Institute of Anthropology. The Warsaw project involved over 100 pairs of twins of which 50 were examined longitudinally; the Wroclaw study (since 1975 directed by P. Bergman) ended with growth curves covering a period of 10-11 years. Monograph analysis (part I) is published in *Materialy i Prace Antropologiczne* vol. 110, 1990.

Both projects presented a classical analysis of intra-pair differences in body dimensions, menarcheal age, skeletal and sexual maturity, fatness and somatotype.

7. ANTHROPOLOGISTS IN PHYSICAL EDUCATION

During the post-war period anthropologists have played an influential role in this field — as researchers, academic teachers, consultants and as long-term rectors of the Academies of Physical Education (A. Sklad at Warsaw, Z. Drozdowski at Poznań, S. Panek at Cracow, A. Janusz at Wroclaw).

Research has concerned the following areas: factors of body build and their role as determinants of motor fitness, the influence of physical exercise on growth and body composition, and the effects of maturity, status and physique on working capacity.

8. ENGINEERING ANTHROPOMETRY

This has been an important line of activity of physical anthropologists in the post-war period. Work in this field has proceeded in two directions: (1) application of anthropometry in clothing design and (2) the use of static and functional anthropometry for design of operator work space.

IV. PHYSICAL ANTHROPOLOGY IN POLAND: FUTURE DIRECTIONS

There are changes under way in Poland in the political, social and economic system of the state. There are also changes in the organizational economic system of financing scientific research, structure of higher education, principles of advancement and development of sciences and the model of social and cultural functions of science. In such a situation it is very difficult to assess trends of the future of scientific research in anthropology, as it is a discipline which does not play a decisive role in the program of modernization and reconstruction of the Polish state.

In connection with the change in the principles of financing science (system of grants and financing of institutions of a high level of scientific development), we should expect important changes in the subjects which interest Polish anthropologists.

The main testing grounds for Polish anthropologists are, and will be in future, local populations inhabiting the river basins of the Vistula and the Odra in prehistory, historical times and the contemporary population in Poland. As far as the theoretical, methodological, and hypothetical approaches are concerned, Polish anthropology uses the same outlook as is found in other parts of the world. It only differs in the selection of empirical material and in the area of interest connected with the European continent as well as in the scientific and cultural tradition of Central Europe.

In the 1970s four main trends were formed under the influence of a strong interest by anthropologists in methodological problems, and because of the assumption of the systems approach adopted from general ecology:

1. Social conditions of the biological state of Polish population;
2. Ecological and genetic conditions of biological transformations of human populations;
3. Biocultural evolution of man;
4. Biosocial conditions of physical and social activity of man.

The most active anthropological institutions in Poland are interested in the realization of the above mentioned research programs.

Future directions of anthropological studies in Poland, according to my subjective opinion, will be concentrated mainly on the following problems:

1. Biological value of the population and analysis of the direction of intensity of transformations;
2. The role of social and cultural factors in the formation of the biological state of Polish population;
3. Methodological and methodical foundations in studies of Man's physical development;
4. Ecological and genetic conditions of biological transformations of human populations;
5. Ecology of subfossil and contemporary populations with special attention regarding the role of cultural and social factors in the level of adaptation of human groups;
6. Biocultural evolution of man, and in particular, the role of cultural and social factors in the formation of microevolutionary variability and of adaptive phenomena;
7. Philosophical and ideological significance of anthropology in the study of the generalized picture of the world and its role in the evaluation of peculiarities of the nature of the human species.

REFERENCES

(SELECTED PAPERS ON DIFFERENT PARTS OF POLISH PHYSICAL ANTHROPOLOGY)

HISTORY OF PHYSICAL ANTHROPOLOGY IN POLAND

Bielicki, T., T. Krupiński, J. Strzalko, 1985. History of Physical Anthropology in Poland, *Occasional Papers*, vol. 1, no.6, International Associations of Human Biologists.

Czekanowski, J., 1956. Sto lat antropologii polskiej 1856-1956, Ośrodek Iwowski, *Materialy i Prace Antropologiczne*, vol. 34.

————., 1962. The theoretical assumptions of Polish anthropology and the morphological facts, *Current Anthropology*, 3: 481-494.

Drozdowski, Z., T. Dzierzykray-Rogalski, 1982. Directions of development and trends in the Polish anthropology during the post-war period, (in Polish), *Przegląd Antropologiczny*, vol. 48: 67-86.

————., T. Krupinski, J. Strzalko, 1983. Polish Anthropology '82, (in Polish), *Przegląd Antropologiczny*, vol. 49: 241-261.

————., 1986. Anthropology and applied sciences. Polish anthropology of the coming years (in Polish), *Przegląd Antropologiczny*, 52: 55-68.

————., 1988. Polish anthropology towards the end of the 20th century - directions and trends of development of its social functions, (in Polish), *Przegląd Antropologiczny*, 54: 186-200.

Godycki, M., 1958. Ośrodek poznański, *Materialy i Prace Antropologiczne*, vol. 39.

————., 1967. Bibiografia antropologii polskiej od roku 1956 do 1965 wiacznie, *Przegląd Antropologiczny*, 33: 5-196.

————., 1973. Bibiografia antropologii polskiej (1966-1970), *Przegląd Antropologiczny*, 39: 1-130.

————., 1976. Bibiografia antropologii polskiej (1971-1975), *Przegląd Antropologiczny*, 42: 3-130.

————., 1982. Bibiografia antropologii polskiej (1976-1980), *Przegląd Antropologiczny*, 47: 3-142.

Jasicki, B., 1957. Sto lat antropologii polskiej 1856-1956, Ośrodek krakowski w latach 1908-1956, *Materialy i Prace Antropologiczne*, vol. 33.

Malinowski, A., N. Wolanski, 1985. Anthropology in Poland, (in:) *Theory and empiricism in the Polish Anthropological School*, (eds) J. Piontek, A.

Malinowski, Seria antropologia nr 11: Poznań.

————., 1986. The sixtieth anniversary of the Polish Anthropological Society (in Polish), *Przegląd Antropologiczny*, 52: 7-14.

Piontek, J., A. Malinowski (eds), 1985. Theory and empiricism in the Polish Anthropological School, *Seria antropologa nr 11*, Poznań.

Reicher, M., W. Sylwanowicz, 1956. Sto lat antropologii polskiej 1856-1956, Osrodek wilenski, *Materialy i Prace Antropologiczne*, vol. 38.

Wrzosek, A., 1956. Bibiografia antropologii polskie do 1956 roku wlacznie, *Materialy i Prace Antropologiczne*, vol. 41.

ANTHROPOGENESIS AND HUMAN EVOLUTION

Bielicki, T., 1985. On a certain generic pecularity of man, *Journal of Human Evolution*, 14: 411-415.

Fialkowski, K.R., 1990. An evolutionary Mechanism for the Origin of Moral Norma: Towards the Meta-Trait of Cultural, *Studies in Physical Anthropology*, 10: 149-164.

————., 1990. An Evolutionary Mechanism for the Origin of Moral Norms; Towards the Meta-Trait of Culture, *Human Evolution*, 5: 153-166.

Góralski, A., A. Wierciński, 1964. An attempt to formalize a concept of the phylo- and ontogeny of the human skulls, *Report of the VI-th Internat. Congress on Quaternary. Warszawa 1961*, 159-290.

Halaczek, B., 1985. Hipotetyczne elementy w przyrodniczej nauce o powstaniu i zaczatkach czlowieka, *Collec. Theol.*, 4: 5-13.

Henneberg, M, 1986. Human cranial capacity decrease in Holocene: A result of generalized structural reduction, *American Journal of Physical Anthropology*, 69: 213-214.

Kaszycka, K., 1984. Zróznicowanie plioplejstocenńskich Homonidea, part I. Charakterystyka morfologiczna i ekologiczna, (Differentiation of Plio-Pleistocene Hominids I. Morphological and ecological characteristics), *Prezgląd Antropologiczny*, 50: 277-297.

————., 1985. Zróznicowanie plio-pleistocenskich Homonidae. II. Dymorfizm plciowy (Differentiation of Plio-Pleistocene Hominids. II. Sexual ddimirphism), *Prezgląd Antropologiczny*, 51: 65-78.

————., 1986. Zroznicowanie plio-plejstocenskich Hominidea. Part III. Konsekwencje taksonomiczne i antropogenetyczne (Differentiation of Plio-Pleistocene Hominids. PartIII. Taxonomic and anthropogenetic consequences), *Prezgląd Antropologiczny*, 52: 129-150.

Klawiter, A., 1991. Adaptation and some its forms, *Variability and Evolution*, 1: 7-24.

Krupiński, T., Z. Rajchel, 1973. Proba otworzenia czaszki *Oreopithecus bambolii Gervais*, *Prezgląd Zoologiczny*, 17: 98.

————., Z.Rajchel, 1976. Proba rekonstrukcji czaszki *Gigantopithecus blacki*

I, Materialy i *Prace Antropologiczne*, 92: 3.

—————., Z. Rajchel, 1987. Odtworzenie glowy *Oreopithecus bambolii Gervais*, *Prezgląd Zoologiczny*, 18: 2-8.

—————., Z. Rajchel, 1985. Odtworzenie czaszki *Gigantopithecus blacki* III — przedstawiciela koplnych Hominidea, *Materialy i Prace Antropologiczne*, 106: 59-65.

., Z. Rajchel, 1987. Odtworzenie glowy *Oreopithecus bambolii Gervais*, *Acta Universitatis Wratislaviensis*, 926: 13.

Kunicki-Goldfinger, W.J.H., 1986 Pozycja *Homo sapiens* seen from the molecular perspective), *Przegląd Antropologiczny*, 52: 15-34.

Lastowski, K., 1991. Two models of evolution in Darwin's theory, *Variability and Evolution*, 1: 25-38.

Rajchel, Z., 1988. Antropologiczna rekonstrukcja czesci kostnych i miekkic glow form kopalnych Hominoidea, Acta Universitatis Wratisl., *Przegląd Zoologiczny*, 20: 1-91.

—————., 1990. General remarks on anthropological reconstructions and the practice of skull and head reconstructions in Poland, *Studies in Physical Anthropology*, 10: 3-67.

Rosiński, F., 1986. Antropogeneza a teologia (Anthropogenesis and Theology), *Przegląd Antropologiczny*, 52: 81-88.

Steślicka, W., 1961. Anthropological investigations of skiers, *The Mankind Quarterly*, 2, 85-91.

Wiercinski, A. 1978. Ontogenetic retardation and human evolution, in: *Proc. Symp. Natural Selection CSAV*, Praha, 271-301.

Zaborowska, B., 1987. Wybiorczosc pokarmowa pawoanow masajskich (Papio cynocephalus) w warunkack wroclawskiego ogrodu zoologicznego (Food selectivity by *Papio cynocephalus* in the conditions of Wroclaw Zoo), *Przegląd Antropologiczny*, 53: 163-170.

HUMAN GROWTH AND DEVELOPMENT

Bielicki, T., H. Szczotka, J. Charzewski, 1981. The influence of three socio-economic factors on body height in Polish Military Conscripts, *Human Biology*, 53, 543-555.

—————., A. Waliszko, B. Hulanicka, K. Kotlarz, 1986. Social-class gradients in menarcheal age in Poland, *Annals of Human Biology*, 13: 1-11.

—————., Z. Welon, 1982. Growth data as indicators of social inequalities the case of Poland, *Yearbook of Physical Anthropology*, 25: 153-167.

Brajczewski, Cz., 1988. Socio-economic differences in body build and work capacity in young men, *Studies in Physical Anthropology* 9: 67-93.

—————., 1990. Ontogenetic and secular changes in the traits of the head in adult males, *Studies in Physical Anthropology* 10: 113-139.

Cieślik, J., M. Sitek, 1987. Selection of the best average normal populations (in Polish), *Przegląd Antropologiczny*, 53: 35-50.

Hulanicka, B., C. Brajczewski, W. Jedlinska, T. Slawińska, A. Waliszko, 1990. City-Town-Village: *Growth of children in Poland in 1988*, Monographies of the Institute of Anthropology Polish Academy of Sciences, Wroclaw.

Jedlińska, W., T. Slawinska, 1990. Influence of family socio-economic status on body height in children from a poor farming region, *Studies in Physical Anthropology*, 10: 69-89.

————., T. Slawinska, D. Kotlarz, 1988. Influence of family socio-economic status on body height in children from a rich farming region, *Studies in Physical Anthropology*, 9: 17-37.

Koniarek, 1988. The skeletal development in twins, *Materialy i Prace Antropologiczne*, 108: 273-285.

Malinowski, A., 1986. Conceptions of norm and normality in the biology of Man and in medicine, *Studies in Human Ecology*, 7, 7-31.

————., M. Pezacka, 1991. Factors conditioning the level of physical development of children in Wielkopolska in 1980-1982, *Variability and Evolution*, 1: 113-125.

Orczykowska-Swiatkowska, Z., B. Hulanicka and K. Kotlarz, 1988. Intrapair differences in somatotypes in three phases of ontogenetic development of twins, *Studies in Physical Anthropology*, 9: 39-59.

Slklad, M., 1975. The genetic determination of the rate of learning of motor skills, *Studies in Physical Anthropology*, 1: 3-19.

————., 1977. The rate of growth and maturing of twins, *Acta Gen. Med. Gemell.*, 26: 221-237.

————., 1977. Skeletal maturation in monozygotic and dizygotic twins, *Journal of Human Evolution*, 6: 145-149.

Siniarska, A., 1993. Socio-economic conditions of a family and somatic and physiological properties of parents and offspring, *Studies in Human Ecology*, 10 (in press).

Waliszko, A., 1988. The evolution of social gradients in menarcheal age in Wroclaw between 1966 and 1976, *Studies in Physical Anthropology*, 9: 3-15.

————., W. Jedlińska, 1976. Wroclaw Growth Study. Part 2: males, *Studies in Physical Anthropology*, 3: 27-48.

————., W. Jedlińska, K. Kotlarz, A. Krajewska, T. Slawińska, A. Szwedzinska, 1985. Growth and development of Polish school children examined in 1978, *Studies in Physical Anthropology.*, 8 (3-26).

Welon, Z. 1990. Evaluation of child's physical development, *Studies in Physical Anthropology*, 10: 91-100.

Wierciński, A., 1978. Ontogenetic retardation and human evolution, In: *Proc. Symp. Natural Selection CSAV*, Praha, 271-301.

Wolański, N., 1988. Ecological aspects of the growth and development of Man, *Collegium Antropologicum*, 12 (1): 7-21.

————., H. Chrząstek-Spruch, A. Kozlowska, A. Teter, A.Siniarska, 1988.

The role of culture, living conditions and genes in the growth of 11-year-old children from Lublin, *Actas do 5 Congreso de Sociedade Europea de Antropologia*, vol. 1-Lisboa.

————., L. Januszko, 1986-87. Genes, constitution and culture versus fertility and survival in man, *Antropologia Portuguesa*, 4-5: 159-171.

BIOLOGY OF HUMAN POPULATIONS
(PREHISTORIC, HISTORICAL AND MODERN)

Budńik, A., Przybyszewska M., 1991. Genetic description of the population of Wielkie Drogi villege — a specific case of meting distances distribution, *Variability and Evolution*, 1: 95-104.

Czekanowski, J., 1954. Die Schweizevische anthropologische Aufnahme im Lichte der polnischen Untersuchungsmethoden, *Przegląd Antropologiczny*, 20: 218-314.

Dzierzykray-Rogalski, T., 1980. Paleopathology of the Ptolemaic inhabitants of Dakhleh Oasis (Egypt), *J. Hum. Evol.*, 9: 71-74.

Gladykowska-Rzeczycka, J., 1974. Anthropological investigations on the bone remains from crematoria cemeteries in Poland, *Homo*, 25: 96-116.

Henneberg, M., 1975. Notes on the reproduction possibilities of human prehistoric populations, *Przegląd Antropologiczny*, 41: 75-89.

————., 1976. Reproductive possibilities and estimations of the biological dynamics of earlier human populations, *J. Hum. Evol.*, 5: 41-48.

————., J. Piontek, 1975. Biological state index of human groups, *Przegląd Antropologizny*, 41: 191-203.

Strzalko, J., J. Piontek, A. Malinowski, 1973. Theoretical and methodical foundations of examinations of bones from cremation graves (in Polish), *Materialy i Prace Antropologiczne*, 85: 179-200.

Wierciński, A., 1977. An Anthropological Concept of Culture and Cultural Evolution, in: *Visions of Man and Society in the Scientific Theories and Investigations*, PWN, Warszawa.

Wierciński, A., 1978. The Meaning and Scope of Anthropology, *Collegium Antropologicum*, 2: 10-16.

————., A. Wiercinski, 1978. An Anthropological contribution to the Origin of Slavs, *Collegium Antropologicum*, 2: 148-153.

————., 1980. Individual typology and the intraspecific taxonomy of man, *Przeglad Antropologiczny*, 46: 279-296.

————., 1986. The meaning of adaptive function, In: *Adaptation, Behaviour and Evolution*, (eds.) V.J.A. Novok, V. Vančata and M.A. Vančatovo, *Czechoslovak Academy of Sciences*, 241-247.

————., 1988. Anthropological concept of the human philosophy of live (in polish). In: Essays in General Anthropology, (ed.) J. Piontek, Adam Mickiewicz University Press, *Anthropology* 12: 29-42.

Wolański, N., 1982. Environmental changes and biological status of human populations, *Collegium Antropologicum*, 6 (1): 69-80.

————., 1984. Positive inolices of health, *Acta. Medica Auxologica*, 17: 227-233.

————., 1985. Secular trend, secular changes or long term adaptational fluctuations, *Acta Medica Auxologica*, 17: 7-19.

————., 1988. Health: An ecological problem, *Human Ecology: Research and Applications*, ed. by R.J. Borden, J. Jacobs, G.L. Young, 243-258, Society for Human Ecology. College Park.

————., 1989. Human life and culture: dynamic components of ecosystems, *Zygon*, 24 (4): 401-427.

————., 1990. Origin and methodology of human ecology, *Journal of Human Ecology* (India), 1 (2): 109-119.

————., E. Kowalczyk, A. Teter, 1991. Subsutaneous Fat Tissue Patterning in Polish Populations, *International Journal of Anthropology*, 6: 137-152.

————., A. Siniarska, 1978. *Ecology of nutrition*, Ossolineum, Wroclaw.

————., A. Siniarska, 1982. *Ecology of Human Populations*, Ossolineum, Wroclaw.

————., S.L. Malik, 1979. Modern environment and future of Man, *Acta Antropogenetica* (India), 3 (3-4): 157-162, reprinted in 1984 under title: Human Ecology: the need for a new emerging science, *Studies in Human Ecology*, 5: 7-13.

————., K. Tomonari, L. Januszko, V. Liocheva, S. Chung, S. Tsushima, 1988. Socio-economic and biological factors of families from Poland, Japan, South Korea and Bulgaria, *Collegium Antropologicum*, 12 (1): 87-93.

EPILOGUE:
FUTURE PROSPECTS

We held a round-table discussion on the last day of the symposium. What had we accomplished, and what could we say about the future of the Science of Man in post-socialist Europe? The following remarks were repeated again and again by the participants, and I will try to summarize them.

What will be the outlets for publication of our research? Many of our journals have been discontinued, or they do not have enough funds to handle the research articles being sent them. How much of our work can be absorbed by the journals of the West? Will not the Western scientists themselves eventually become put off by our sending articles for publication in their journals? Will not they expect us to publish primarily in journals of our country, as we would expect them to do?

How much will Western scientists want to come to our countries, now that access is much easier, and study our materials? What will happen to our own research if a large number of scientists from outside, travelling with large grants and more sophisticated equipment, want to study the data that we have accumulated and materials that we have discovered? How much of this will be a gain, and how much a loss?

We still do not have access to the financial support that Westerners do. Westerners have the greater opportunity to travel in carrying out their studies, whether the research be primate studies (which in the wild has to be in the warmer countries) or research based on making comparative studies of original materials located only at specific places. Can we compete in the same fields with others who have more money, more equipement and more sophisticated techniques?

As the borders of our countries alter, the access that we ourselves have to research, also changes. For example, Russian primatologists who now wish to study at Sukumi Primate Research Centre must now travel to a foreign country since the primate facility is located in Georgia, which became a separate country when the USSR was dissolved in 1992. Similar situations could be envisioned for the newly split Czechoslovakia. East German scientists are being assimilated into the united Germany, but is it always in a better circumstance?

On the brighter side, there are a number of advantages that we can see in our future. There is the likelihood of the appearance of new approaches and new methods as a consequence of closer contacts with the West. Liaisons with Western colleagues will generate new types of investigation into the biology of humans. Greater exchange of ideas and data will become possible in which both sides will most likely reap the benefits.

Finally, the greater freedom that the West has had in exploring studies in the Science of Man will become a part of our own everday experience again.

APPENDIX

List of Participants•, Addresses* and Paper Titles for the *Symposium on Foundations for Different Approaches to the Study of Human Evolution,* September 1-3, 1989, Liblice, Czechoslovakia. (Chechoslovak Academy of Sciences.

Co-Chairmen: **B.A. Sigmon**, Department of Anthropology, University of Toronto, Toronto, Canada.
V. Leonovičová, Department of Evolutionary Biology, Czechoslovak Academy of Science, Prague, CSSR.

N. Bonde and P. Bennike
Institute of Historical Geology and Paleontology
University of Copenhagen, Oster Voldgrade 10
1350 Copenhagen K, Denmark
Physical Anthropology and Human Evolutionary Stages in the Historical Development of Studies in Denmark and Other Scandanavian Countries.

T. Bromage
Department of Anthropology
Hunter College, C.U.N.Y
695 Park Ave.
New York, NY   10021
New Approaches to Human Evolutionary Studies: Exploring Growth Patterns in Fossils

M. L. Butovskaya
Institute of Etnography
Academy of Sciences
ul. Dmitrija Uljanova 19
SU - 117036 Moscow, Russia
Ethological Approach to the Study of Human Evolution

•B.A. Chiarelli
Instituto di Anthropologia
via del Proconsola 12
50122 Firenze, Italyl
The Development of Physical Anthropology / Human Evolutionary Studies in Western Europe Since World War II

Professor O.G. Eiben
Department of Anthropology
Eotvos Lorand University
Pushkin uten, 3
H-1088, Budapest, Hungary
Specific Approaches and Research Directions in Somatology: Implications for Human Evolution

Professor A. Forsten
Zoological Museum
P. Rautatiekatu 13
00100 Helsinki, Finland
Contribution to Evolutionary Research from Non-Hominid Animal Evolutionary Studies. Equids.

•A. Ghosh
35 Ballygunge Circular Rd.
Calcutta 29
West Bengal, India
Physical Anthropology / Human Evolutionary Research From the Indian Perspective

K. Hanihara
International Research Centre for Japanese Studies
Rakusai Centre Bldg.
2-5-9 Higashi Sakaidani-cho, Oharano, Nishikyo-ku, Kyoto 610-11, Japan
Japan and Human Evolutionary Research

V. Leonovičová
Department of Evolutionary Biology
Czechoslovak Academy of Sciences
Na Folimance 5
CS - 120 00, Prague 2, CASSR

Sociobiological Perspectives in Human
Evolutionary Research

J. Linhart
Institute of Psychology
Czechoslovak Academy of Sciences
Prague, CSSR.

The Principle of Reflection in Human Research

S. Løvtrup
8 Rue Buffon
Museum National d'Histoire Naturelle
Paris, France

Human Evolution and Epigenetics

•F.J. Melbye
Department of Anthropology
University of Toronto
Toronto, Canada

Studies of Human Evolution / Physical
Anthropology in Canada

V. Novak
Department of Evolutionary Biology
Czechoslovak Academy of Sciences
Na Folimance 5, CS-120 00
Prague 2, CSSR.

Results of Human Evolutionary Research in
Czechoslovakia

V. Novotny
Laboratory of Evolutionary Biology
Czechoslovak Academy of Sciences
Na Folimance 11, CS-120 00
Prague, CSSR.

Systems Approach in Human Evolutionary
Research

C. Oxnard
Department of Anatomy and Human
Biology
University of Western Australia
Nedlands, W. Australia

Anatomies and Lifestyles, Morphometrics and
Niche Metrics: Tool for Studying Primate
Evolution

V. Piesce-Delfino
University de Bari
Instituto di Anatomia e Istologia
patologica
Policlinico, I-70124 Bari, Italy

Different Approaches in Shape Understanding
in Anthropology

Janusz Piontek
Department of Anthropology
Adam Mickiewicz University
ul. Fredry 10, PL-61-701
Poznań, Poland

The Problem of "norm" in Human Evolutionary
Biology

W.D. Ross
Department of Kinesiology
Simon Fraser University
Vancouver, BC   V5A 1S6

Iconometrographics in Human Evolutionary
Biology

A. Santangelo
p. Giov. Bande Nere
2 20146 Milano, Italy

The Beginning of Culture

S. Saunders
Department of Anthropology
McMaster University
Hamilton, Ontario
L8S 4L8

A Review of Evolutionary Theorists and Their
Influence in North America and Europe

Ecology: The Development of the Relation
Between Man and Nature

V. Sedivy
Vue Zpok
Botanicka 68 a 656 01
Brno, CSSR

B. Senut
Laboratoire D'Anthropologie Biologique
Musée de L'Homme
17, place du Trocadéro
75116 Paris, France

French Contributions to the Study of Human
Evolution

B.A. Sigmon
Department of Anthropology
University of Toronto
Toronto, Ontario   M5S 1A1
Canada

Theoretical Models in Human Evolutionary
Studies

Dr. Jaroslav Slípka
301 66 PLZEN
Karlovarska 48
C.S.S.R

Evolution of the Human Immune System

I.N. Smirnov
Leninsy Prospekt
D. 69, 44
Moscow, Russia

The Study of Human Evolution from
Philosophical Perspectives

J. Svoboda
Archaelogicky Ustav CSAV
Sady Osvobozeni 17/19
CS - 662 03 Brno, Czechoslovakia

Paleolithic Evolution in Moravia/
Czechoslovakia

Herbert Ullrich
Institut für Anthropologie der
Humboldt-Universität, Invalidenstrasse
43, D 0-1040 Berlin, Germany

Human Evolutionary Research in the German
Democratic Republic

V. Vančata
Department of Evolutionary Biology
Czechoslovak Academy of Science
Na Folimance 5
CS-120 00 Prague 2, CSSR

Evolution of Human Femur and Tibia: A
Morphometric Approach to Human
Evolutionary Research

M.A. Vančatova
Institute of Physiology
Czechoslova Academy of Science
Vidensk 1083
CS - 142 20 Praha 4 - Krc, Czechoslovakia

Relevance of Primate Behavioural Studies for the
Investigation of Human Evolution

•E. Vlček
Narodni muzeum v Praze
115 79 Praha 1, Tr. Vitezneho unora 74
CSSR

The Meaning of Czechoslovak Paleoanthropological Findings for Research in Human
Evolution

A. Wiercińska, A. Wierciński,
M. Wierciński
Lab of Anthropology
State Archeological Museum
Dluga ul. 52, PL-00-950
Warszawa, Poland

Some Results of the Studies on Mesolthic Finds
from Poland. /On the Concepts of
Taxonomically Distinctive Nature and Essence of
Man. /The Analysis of Some Parameters of
Socio-Cultural and Biological Status of Female
Students of Kielce

•W. Rukang
Institute of Vertebrate Paleontology
and Paleoanthropology
Academia Sinica
P.O. Box 643, Beijing (28) China

Human Evolutionary Research: Chinese
Perspectives

N. Yamazaki
Department of Mechanical Engineering
Faculty of Engineering
Keio Univ., 3-14-1, Hiyoshi, Kohoku-Ku
Yokohama 223, Japan

A Biomechanical Relationship Between Body
Proportion and Its Locomotion

•   This indicates those participants
     who wrote papers but were
     unable to attend the symposium
     in person.
*   These are the current addresses
     for 1992.